SOTHIS PRESS

Sothis Press aims to completely reimagine the public domain book with an emphasis on universal access to captivatingly designed print editions. We believe that technology will enable every single book ever published to be perpetually accessible in beautiful print form. Sothis is beauty in ubiquity.

www.sothis.press
1 Waterhouse Square, London, ECIN 2ST

# Cam Design and Manufacture

F. B. Jacobs

SOTHIS PRESS

ISBN 978-1-0368-0955-3

*Cam Design and Manufacture*
is in the public domain.

This edition is published by
Sothis Press © 2023. The company
is registered in England and Wales.
The Company Number is 14900389.

# CAM DESIGN
## AND MANUFACTURE

BY

F. B. JACOBS

*87 ILLUSTRATIONS*



NEW YORK
D. VAN NOSTRAND COMPANY
EIGHT WARREN STREET
1921

Printed in the United States of America

# PREFACE

Cam movements are among the most useful that the machine designer has at his command, for without them many complicated motions could not be laid out and constructed. When cams are laid out by cut-and-try methods (which is often the case), they are noisy in operation, create a vast amount of friction and are short-lived. On the other hand, if the curves are properly laid out, giving all possible time for the rises, the finished cams can be run at high speeds with very little noise or friction. Under these conditions, cams compare favorably in construction cost and maintenance cost with more complicated mechanical movements used for the same purpose.

When it is considered that a practical knowledge of cam design and manufacture is easily acquired, it may seem strange that the majority of technical works on machine design treat this subject lightly; or at least from only a theoretical point of view. While it can not be claimed that any one person knows the whole subject thoroughly, it is certainly possible to explain some of the methods that are in actual use for laying out and cutting cams. The use of the various methods and appliances that are described is not confined to any one locality, by any means, as these methods and appliances have been found in practically every manufacturing center of the country.

In the description of the various methods, the writer has avoided using complicated mathematical formulas as it is well known that very few men actually engaged in machine design and construction thoroughly understand them. A working knowledge of geometry, algebra and trigonometry is, without a doubt, a valuable asset for every man, and it is hoped that the practical men of the future will understand these subjects thoroughly.

At the present time, however, it can not be denied that one reason why technical literature does not find favor with the majority of practical men is because the writers persist in using mathematical formulas that are not understood by their readers.

FRED. B. JACOBS

CLEVELAND, OHIO,
February 15, 1921

# CONTENTS

# CAM DESIGN AND MANUFACTURE

## CHAPTER I

### Machine Cam Design

The majority of cam movements may be described as devices for changing rotary motion into regular or intermittent reciprocating motion. Other forms of cams, are utilized for changing the direction of reciprocating motions. The majority of cams used in machine construction are of the first type, as seen on shoe and textile machinery, printing presses, and other forms of automatic machinery.

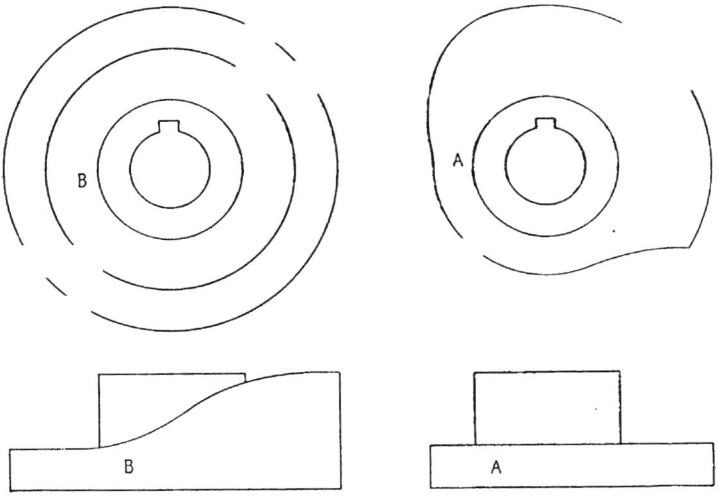


Fig. 1

The two cams illustrated in Fig. 1 are called "open cams" because the roll bears at only one point at a time. "*A*" is a disk cam, the movement being developed on the periphery. "*B*" is a drum cam, in which the movement is developed on one side. These are sometimes called barrel cams. An objection to open cams is that a weight, or spring, has to be employed to keep the roll in position. However, as cams of this type can often be cut without a master they are frequently used for experimental pur-

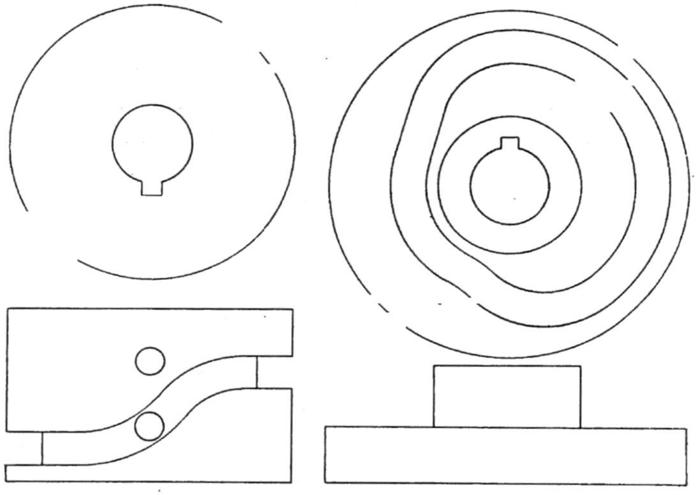

FIG. 2.

poses. Again, they possess the advantage of giving more abrupt rises than is possible with path cams, and for this reason they are used on automatic screw machines, for actuating gas engine valves, and for various other purposes where rapid rises are desired.

The cams shown in Fig. 2 are called "path cams" because their rolls are confined to a path, or groove. They are generally seen in places where accurate movements are required, and where it is not practicable to utilize a weight, or spring, to keep the roll in contact. As these cams are the hardest to lay out and cut, their use is generally confined to accurate machinery. Cams of this type, however, possess one disadvantage; that is their inability to take extremely rapid rises, the reasons for which will be explained later on.

While both open and path cams can be of either type, that is disk or drum, the disk type is the more popular as it is easier to cut, runs with less friction, and has no disadvantages worth mentioning. Drum cams create an unnecessary amount of friction and wear on their rolls, owing to the fact that the outer edge of the path has a greater peripheral speed than the inner, due to the increase in diameter.

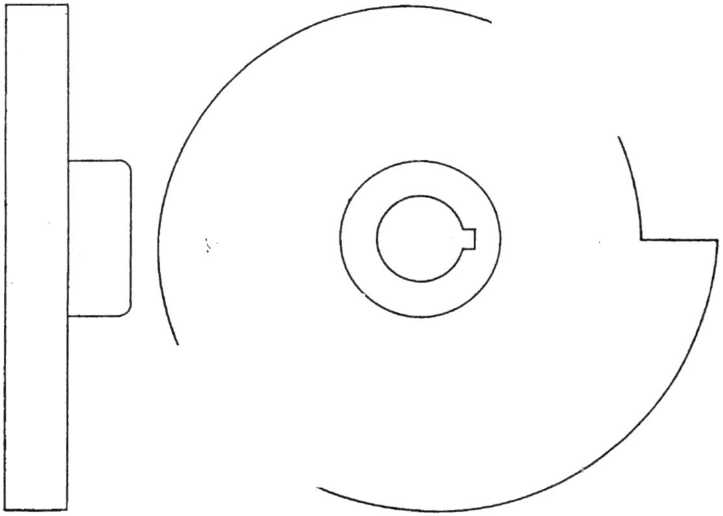

FIG. 3

It can be said in favor of the drum cam that it is very easily designed, as the development is laid out flat, and in straight lines, instead of in converging lines and circles as with disk cams. Again, it is often necessary to employ drum cams in cases where the movement runs parallel with the cam shaft.

In Fig. 3 is shown a form of cam called a wiping cam. These are used in cases where a sudden drop is required.

As the follower is a fixed piece in place of the usual roll, cams of this type are short-lived: a large amount of friction being present, especially in cases where the operator neglects to oil the movement frequently. Notwitstanding this objection, this form of cam is frequently employed as it gives a simple method of acquiring the desired motion.

The cam illustrated in Fig. 4 is called an eccentric cam. It gives an easy motion in cases where advances with intermittent rests are desired. This type of cam is often

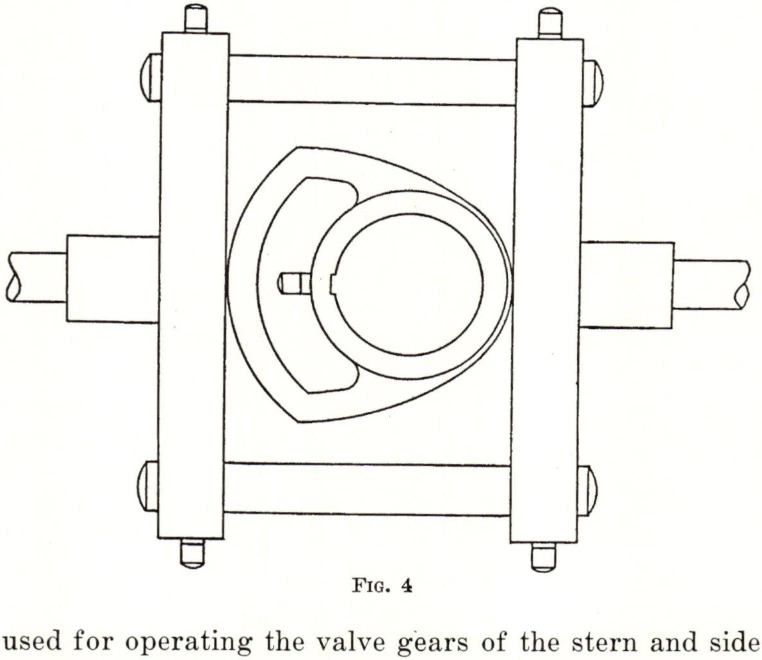

FIG. 4

used for operating the valve gears of the stern and side wheel steamboats that are seen on the Mississippi River and its tributaries.

Eccentric cams are also used in the construction of accurate shoe machinery. When operated at moderate speeds this type of cam works easily, is comparatively long-lived, and calls for little attention, save for an occasional adjustment of the plates that confine it. There are many other forms and modifications of cams, includ-

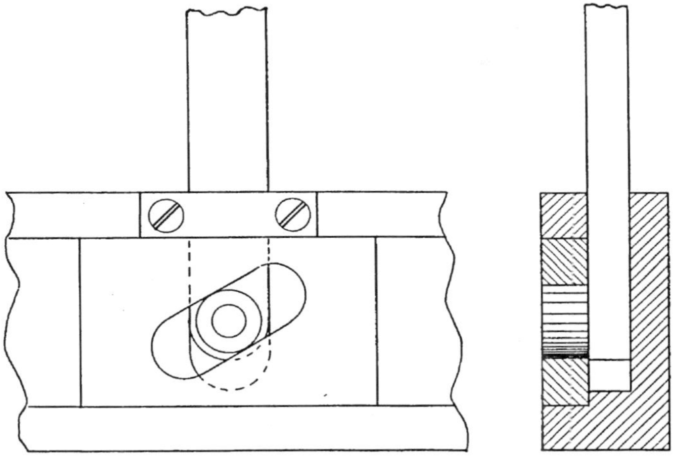

FIG. 5

ing plate cams, one of which is shown in Fig. 5, but the cams above described are the types most frequently used.

The first step in laying out a cam is to consider the motion required with the intervals of action and rest; it being assumed that the type of cam whether open or

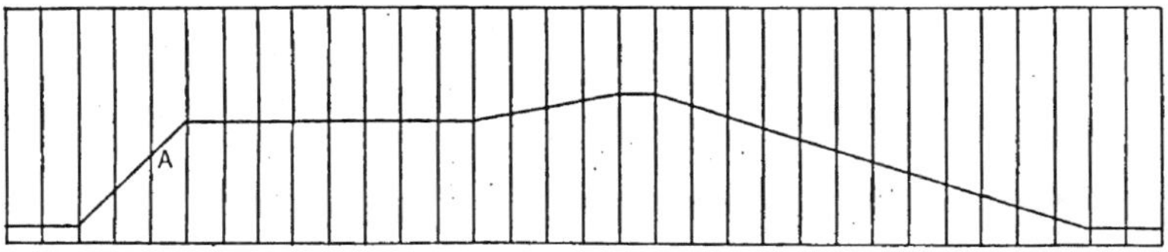


FIG. 6

path, disk or drum, has been decided upon. A chart, as shown in Fig. 6, is now drawn. This shows by its ordinates the rises and rests in their exact proportion as they will appear in the finished cam. If the cam is to be of the drum type the length of the chart should just equal the circumference of the finished cam. If a disk cam is to be used the length of the chart should equal approximately the smallest circumference of travel in the finished cam. This precaution is taken to avoid too abrupt rises which would work hard, and in extreme cases lock.

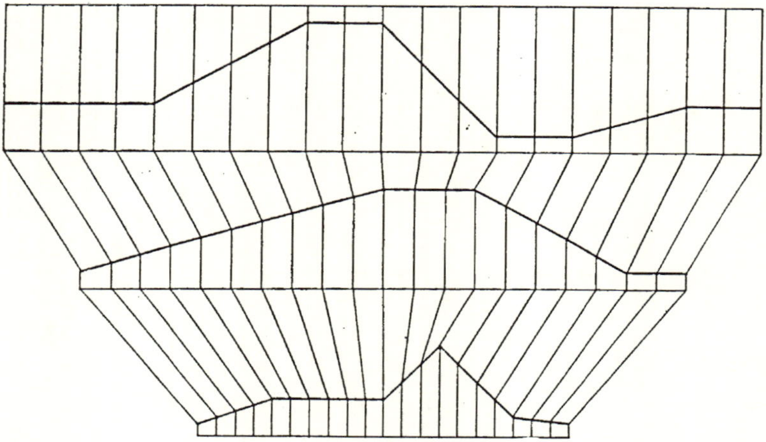

FIG. 7

This is readily understood when we stop to consider that the base and hypothenuse of the triangle, showing the angle of motion, change as the circumference of the cam is made greater or less, while the altitude, which represents the stroke, is always fixed as it represents a definite dimension. Thus, if Fig. 6 represents the circumference of a cam leader, from which a smaller cam is to be cut, which is the generally accepted practice, it is evident that the angle shown at *"A"* would be made more abrupt as the circumference decreased. As it is an angle of 45° at present, which is practically the limit for fast and smooth running, it is evident that a sharper angle would cause the cam to work very hard and under some conditions to lock entirely.

Where several cams are to run on one shaft, or on different shafts at the same speed, it is a good plan to make a compound chart as illustrated in Fig. 7, taking care that the lines representing the several circumference are approximately the correct lengths. In a chart of this kind, the lines representing the divisions of the cam are diverted to pass through corresponding points representing equal degrees or divisions of a circle. By referring to this chart the designer can see before him all the motions for a machine in their correct relation to each other. In many cases, this saves confusion in trying to compare one cam motion with another, laid out on a different chart. From the cam chart the leaders are laid out, but before taking up this point it may be well to consider a few of the cam curves, which represent the travel of the center of the cam roll, and a difficulty, often met with, called "cutting back."

There are three varieties of cam curves in common use. These are shown in Fig. 8. Illustration *"A"* is called the uniform motion or regular rise and is used for converting rotary motion into regular reciprocating motion. This is a hard working movement as it starts and comes to a rest with a decided shock, even when run at moderate speeds. For this reason it should be avoided as much as possible. As this motion is easily laid out, either for a drum or disk cam, it is often used by inexperienced designers in cases where another motion would serve the purpose to better advantage.

In laying out this motion the base of the triangle equals the division of the cam that is allowed for completing the stroke. The altitude equals the stroke of the cam. The angle, which is not considered unless it is steep enough to be detrimental to smooth running, is determined by the altitude and time division of the cam circle. It is necessary to draw curves at the start and stop of this motion if the cam is to be of the path type.

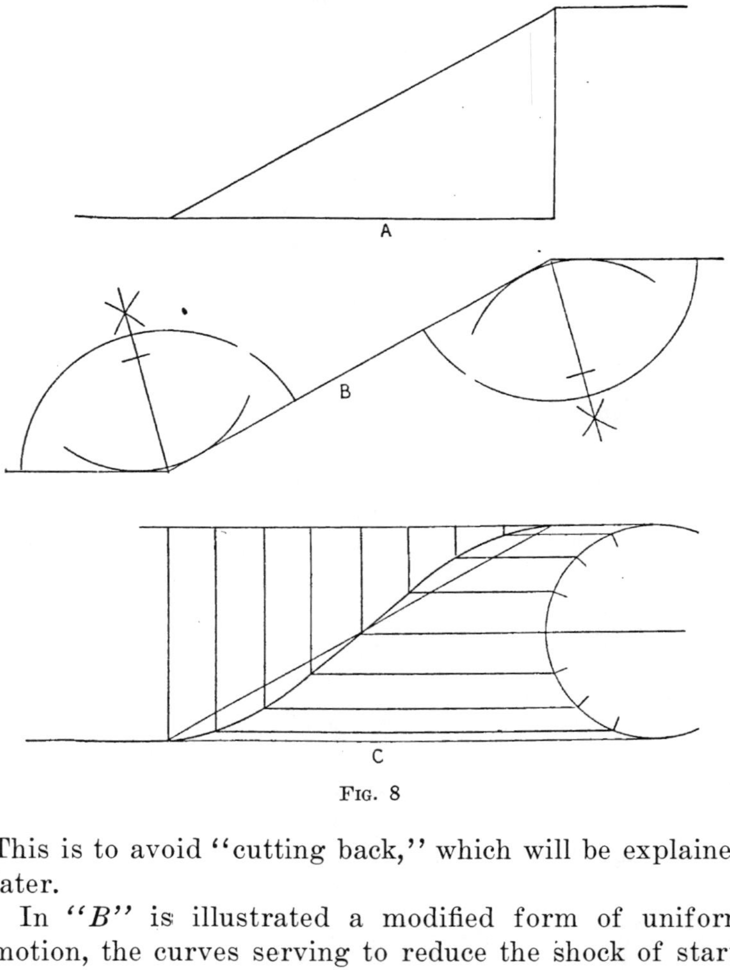


Fig. 8

This is to avoid "cutting back," which will be explained later.

In *"B"* is illustrated a modified form of uniform motion, the curves serving to reduce the shock of starting and stopping. This motion can be used in cases where it is necessary to have a uniform motion for the greater part of the stroke, but where a little time can be spared for getting under way and stopping. When possible, the radii of the curves that fill in the ends should equal the diameter of the cam roll. Many designers claim that this form of curve, which is an inclined plane with the ends modified, is theoretically correct because an inclined plane is one of the elementary mechanical powers put to actual use. Experience, however, has proven that other forms of curves are more quiet in operation and longer lived. In cases where a uniform motion is called for, this curve is an improvement over the direct motion. This curve is drawn, as illustrated, by bisecting the angles and filling in with curves whose radii equal the diameter of the roll.

The crank motion, or harmonic curve, is shown at *"C."* It is a well known fact that any machine part that is actuated by a crank and connecting rod starts and stops slowly, due to the angularity of the rod. The motion in question, which is a modification of the above principle, starts and stops without shock; the greater amount of work being performed after the motion is well under way. This motion can generally be used to advantage, for in the majority of cases it is not necessary that the parts actuated by the cam move with an absolute uniform motion.

This curve is laid out by drawing a circle, the diameter of which is equal to the rise, and dividing it into any convenient number of equal parts. The line equaling the motion of the cam is then divided into the same number of equal parts. The curve passes through the intersections of the lines drawn from the motion line and the circle.

Another motion, called the gravity curve, is sometimes employed. It is developed in the same way as the harmonic curve from an ellipse in place of the circle. The major and minor axes have a relation of 11 to 8, the minor axis representing the stroke of the cam. In actual practice, however, it has never been definitely proven

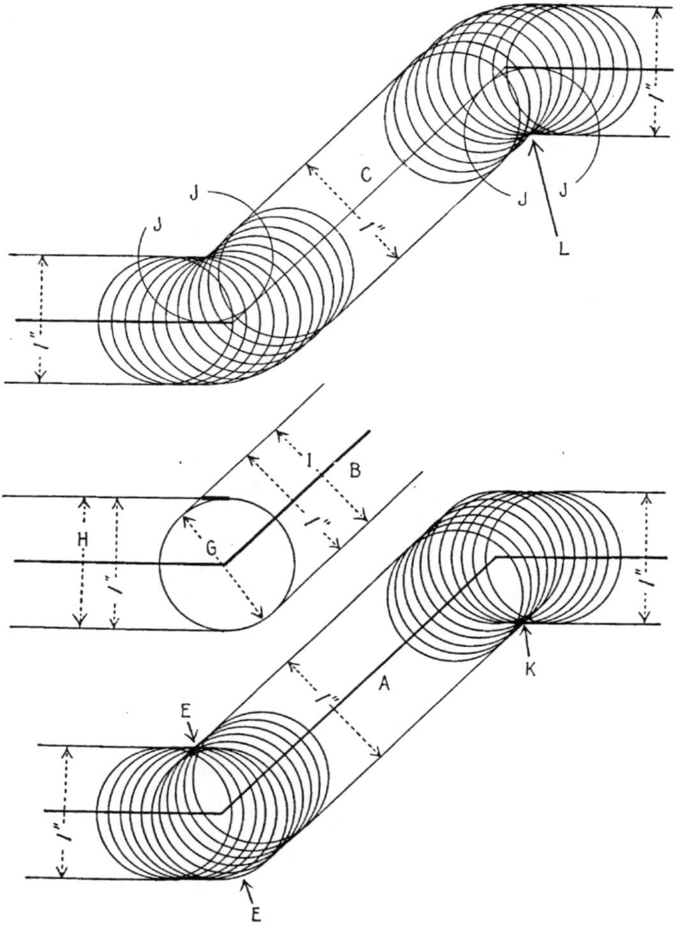


FIG. 9

that the gravity curve works any easier, or is productive of longer life, than the regular harmonic motion. For this reason, together with the fact that it calls for more attention in laying out, both on the drawing board and on the cam leader, it is seldom used in practice.

By referring to "*A*," Fig. 9, it is seen that the distance between "*E E*" is greater than the width of the cam path. In shop language, this is termed "cutting back" and the principle is easily understood by glancing at illustration "*B*," in which "*G*" represents the outline of the milling center that has just completed the path "*H*." In cutting the path at "*I*" the cutter will cut away part of the wall shown by the heavy line, thus making it impossible for the roll to fill the path completely at this point.

This difficulty is overcome by drawing the circle "*J J*" in illustration "*C*," with a radius equal to that of the roll. The center of this milling cutter follows that part of this circle "*J J*" included between the points of tangency with the straight parts of its path. *H*ere it is seen that the roll completely fills the path at all points. It is not necessary to lay out the circle "*JJ*" to develop the point "L," as the cut back is automatically avoided at this point, all the circles shown in "*A*" coming to the point "*K*" just as they do at "L" in illustration "*C*" where the roll circles are drawn from points on the circle "*J J*" between its points of tangency with the straight lines.

With open cams, cutting back does not have to be taken into consideration, as it is only necessary for the roll to have a bearing point on one side. This is the reason why it is possible to obtain very abrupt rises with open cams.

Cutting back is always to be considered where very sharp rises are required, and this factor often determines whether the cam shall be a path cam or an open one. This is explained in Fig. 10. Let it be assumed that it is necessary to obtain the rise between the points

*"A A"* is the shortest possible time; which is determined by the revolution time required for a cam to roll over a given point. This may sound like a paradox, but a little

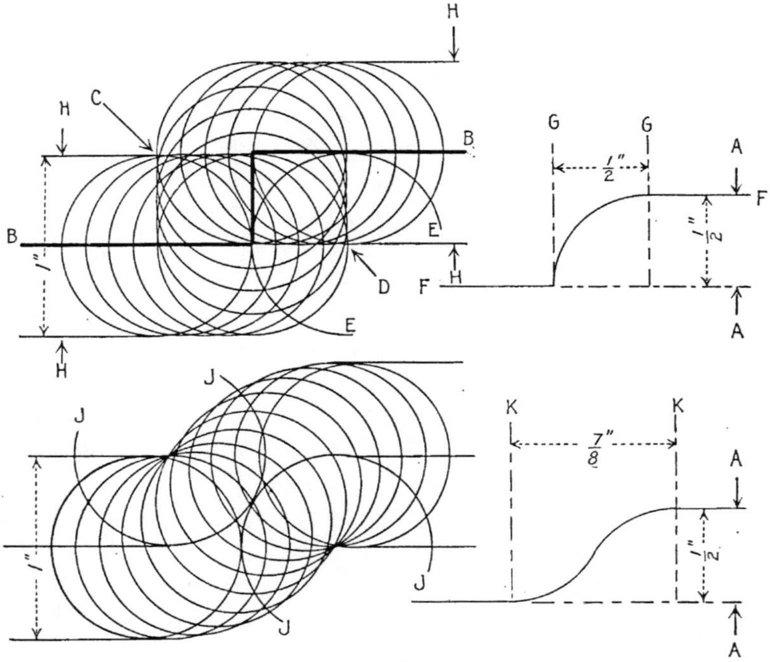


Fig. 10

study of Fig. 10 will impart the theory. Draw the line of motion *"B B."* From points on this line the several circles representing the roll are drawn. As they all come together at the point *"D,"* the center of the roll may travel on the arc *"E E"* thus imparting the motion "F F," in which the rise is accomplished in the time it takes the cam to travel the distance represented by the lines *"G G."* When we measure the distance between the points *"C"* and *"D"* and compare it with the space between the points *"H"* and *"H,"* the width of the cam path, we find that the cut back is very pronounced. Thus, as we desire to obtain the rise in the shortest possible time, this cam would have to be of the open type.

If this cam were to be of the path type, it would be necessary to develop the line of motion on the two arcs *"J J"* to avoid cutting back. In this case, the rise would be accomplished in the time it takes the cam to travel the distance represented by the lines *"K K,"* which is greater than the distance between the lines *"G G,"* as the illustration shows.

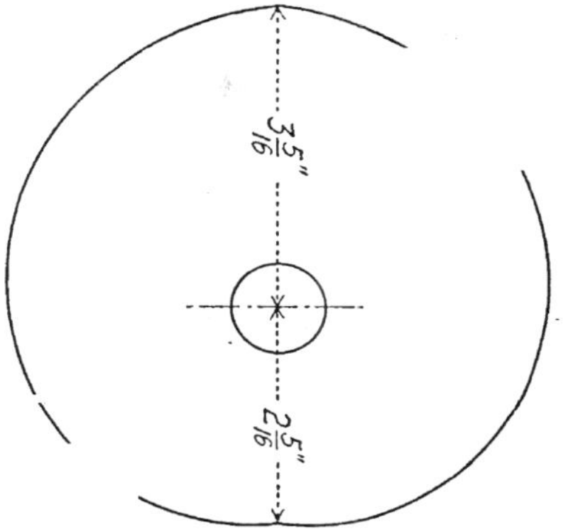

Fig. 11

Before proceeding any farther with the subject, it may be of interest to consider several typical cam movements, taking note of their outlines as they will appear on the finished cams.

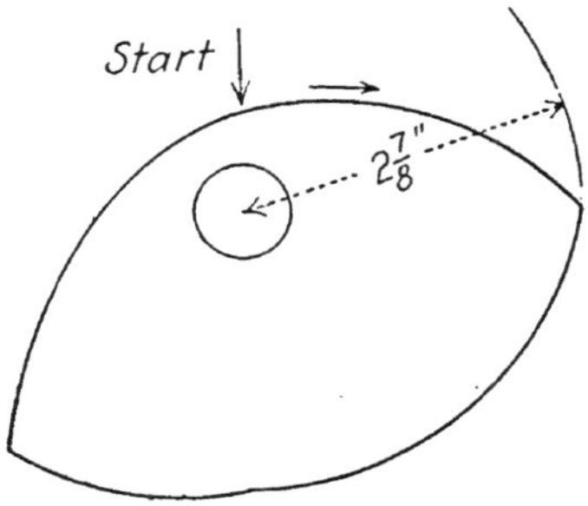

Fig. 12

Fig. 11 illustrates a regular rise and fall movement, often called the "heart-shaped cam," wherein the rise is regular for a half-revolution, followed by a drop wherein the motion of the actuated part is reversed. This motion is used in place of a crank and connecting rod, in cases where errors due to rod angularity would prove detrimental.

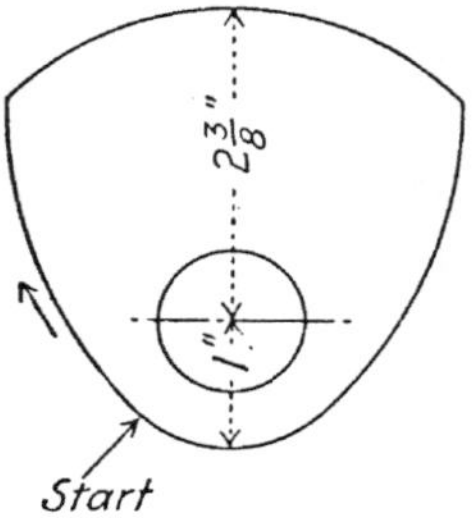


FIG. 13

In Fig. 12 is shown a rise of 2″ in $\frac{1}{4}$ rev., followed by a drop of $\frac{1}{2}$″ in $\frac{1}{4}$ rev., next a rise of $\frac{1}{4}$″ in $\frac{1}{8}$ rev., then a drop of $1\frac{3}{4}$″ in $\frac{3}{8}$ rev., which completes the circle.

Fig. 13 illustrates a rise of $1\frac{3}{8}$″ in $\frac{1}{4}$ rev., a rest of $\frac{1}{4}$ rev., a drop of $1\frac{3}{8}$″ in $\frac{1}{4}$ rev., and a rest of $\frac{1}{4}$ rev. This is an eccentric cam and can be used either with a fixed follower or a roll. It is sometimes called a four-throw eccentric, because if placed in a box that would confine it on all sides it would impart four movements. It is symmetrical about the vertical center line.

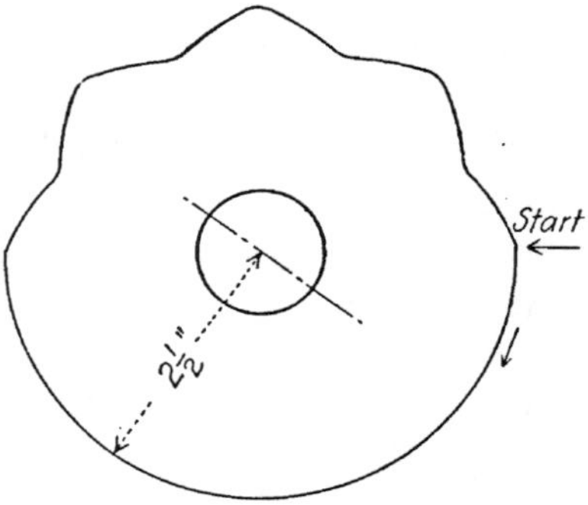


FIG. 14

The movement shown in Fig. 14 consists of a rest of $\frac{1}{2}$ rev., followed by four alternate drops and rises of $\frac{3}{8}$″ in periods of $\frac{1}{8}$ rev.

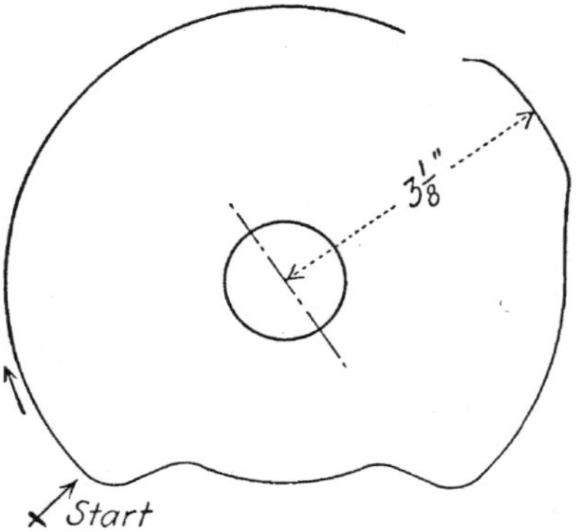


FIG. 15

The motion shown in Fig. 15 illustrates a rest of $\frac{7}{16}$ rev., a rise of $\frac{1}{4}$″ in $\frac{1}{16}$ rev., a rest of $\frac{1}{8}$ rev., a drop of $\frac{1}{4}$″ in $\frac{1}{16}$ rev., a rest of $\frac{1}{8}$ rev., a drop of $\frac{3}{4}$″ in $\frac{1}{16}$ rev., a rest of $\frac{1}{8}$ rev., and a rise of $\frac{3}{4}$″ in $\frac{1}{16}$ rev.

In Fig. 16 is shown a drop of $\frac{3}{4}$″ in $\frac{1}{16}$ rev., followed by a rest of $\frac{1}{64}$ rev., then a rise of $\frac{3}{4}''$ in $\frac{1}{16}$ rev., and a rest for the remainder of the circle. Cams of this kind are sometimes used for actuating feed movements. Fig. 17 shows the same movement reversed. As the sole object is to move a lever a given distance in a given time, either cam will answer the purpose.

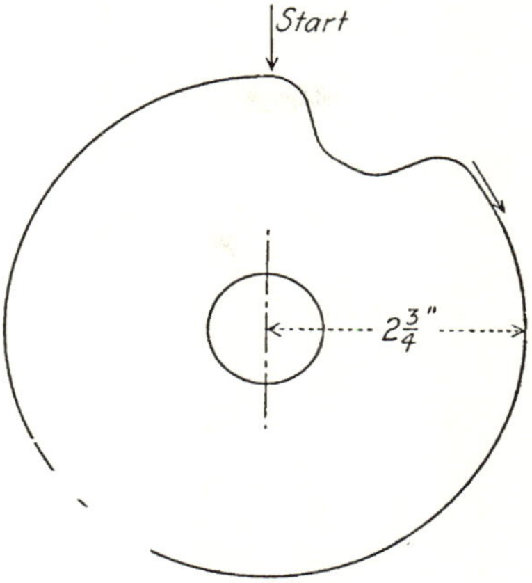


FIG. 16

Fig. 18 illustrates a wiping cam having a sudden drop of $1\frac{1}{2}''$ followed by a rise of the same distance in $\frac{3}{8}$ rev. The remainder of the circle is a rest period. Wiping cams are generally "under-cut" as the illustration shows. This is to give the follower a free movement without danger of cramping.

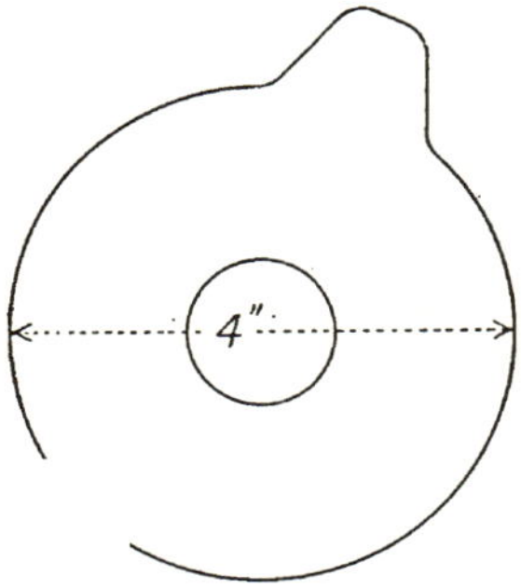


FIG. 17

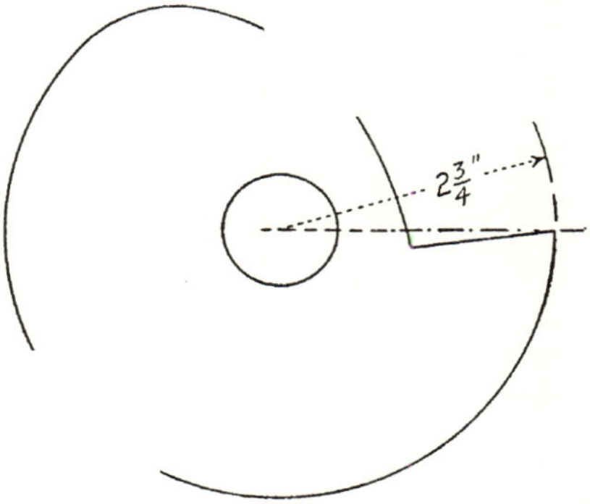


FIG. 18

A cam rocker, which is another form of wiping cam, is shown in Fig. 19. These are used for various purposes, the one illustrated being of the type used for lifting the

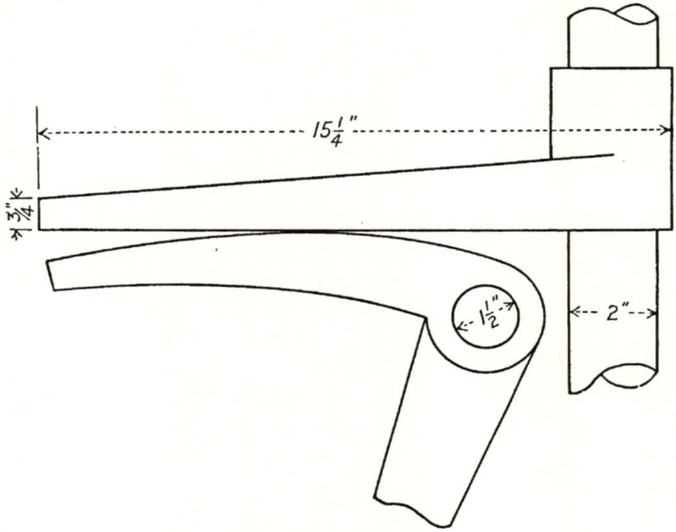


FIG. 19

valves of the beam engines used on side-wheel steam boats. These cams are generally laid out to impart a

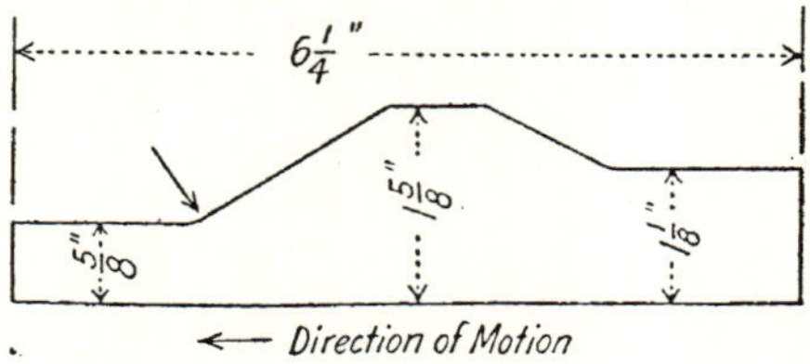


FIG. 20

uniform rise to the follower, which is a flat toe piece, as the illustration shows. The method of generating this curve is explained in Chapter Seven.

ig. 20 illustrates a plate cam. These generally move horizontal or vertical line, although they are somes used in connection with a sector, being actuated by tary motion. The cam shown gives a rise of 1″ dura travel of $1\frac{1}{2}$″, a rest of $\frac{3}{4}$″, a drop of $\frac{1}{2}$″ during a el of 1″ and then a rest. The return stroke is the rse of this movement.

# CHAPTER II

## Gas Engine Cam Design

In the design of cams for operating gas engine valves, it is sometimes the practice to employ the cut-and-try method, which consists of machining and filing an experimental cam until it imparts the desired motion to the valve. More efficient results, however, can be obtained by laying out the cam, as thereby the periods of rest and motion can be exactly proportioned. This is quite necessary as the cam shaft runs only half as fast as the crank shaft, and this requires close adjustment of the periods of rest and action.

In theory, the exhaust valve should open quickly at the completion of the impulse stroke, and close quickly at the completion of the exhaust stroke. In actual practice, however, this is impossible, as the comparative slow speed of the half-time shaft would necessitate steep cam rises, which would cause the cams to be noisy and short-lived.

The difficulty is overcome by starting to open the exhaust valve before the completion of the impulse stroke, and closing it on, or after, the completion of the exhaust stroke. On this same principle allowances are also made for the opening and closing of the inlet valve. The design of cams for the comparatively slow running, horizontal type of gas engine calls for practically the same methods used in designing any machine cam. With automobile, airplane and some types of marine engines, however, a different method is employed, and for this reason the two methods will be described separately.

In Fig. 21 is illustrated the first step that should be taken in laying out the cams for a horizontal gas engine. This is an angularity chart, giving the positions of the piston and crank pin for sixteen equal parts of one half-revolution. It is, of course, necessary that the stroke and length of the connecting rod shall have been determined, as changes in either of these factors would give different movements, due to a change in rod angularity. It is not necessary to mark off any more divisions than a few at the beginning and completion of the stroke, although in this case they are all shown with the idea of making the principle more easily understood. It is readily seen that from a chart of this kind the

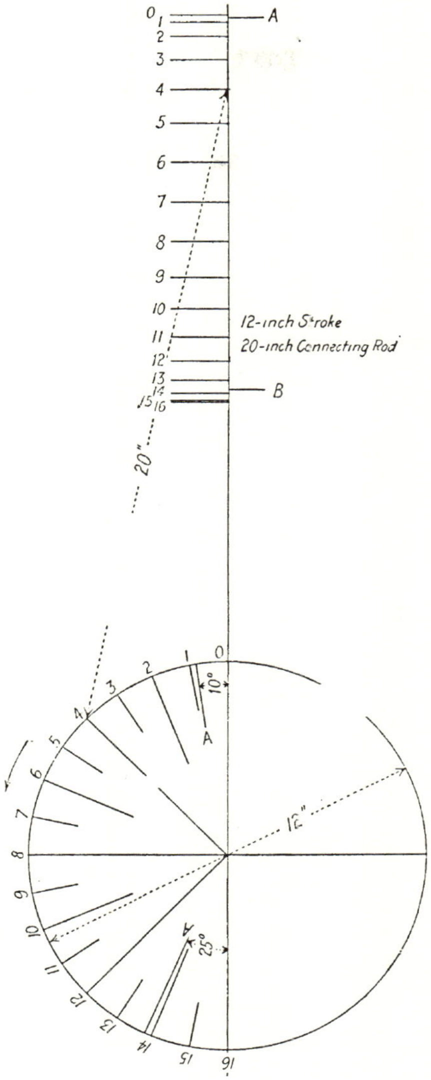


FIG. 21

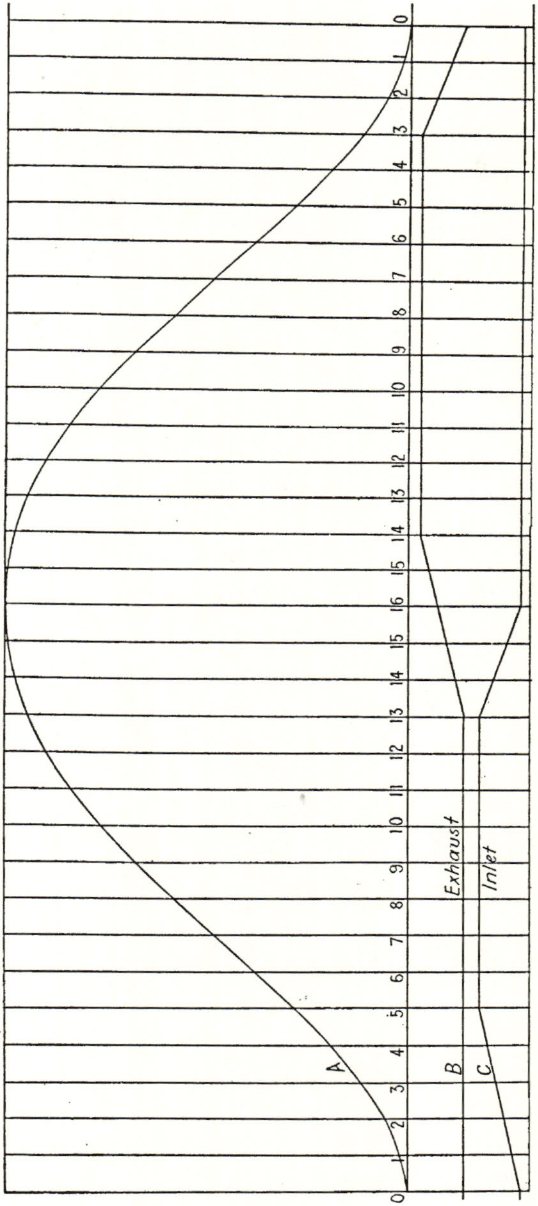


Fig. 22

amount of time that can be spared for the opening and closing of the valves is easily determined. Sixteen divisions are not arbitary, by any means. Some designers prefer to use some factor of 360, thereby representing degrees instead of mere divisions.

A cam chart as shown in Fig. 22 is next drawn. On this chart the line *"A"* gives the motion of the piston for two strokes, the line *"B"* the motion of the exhaust valve, and the line *"C"* the motion of the inlet valve. In connection with the inlet valve, however, the line *"A"* represents the intake and compression cycles, while with respect to the exhaust valve it represents the impulse and exhaust strokes. This eliminates the drawing of the long chart which would otherwise be necessary, as the cam shaft, generally spoken of as the "half-time" shaft, makes but one revolution to two of the crank shaft.

As the cam chart is to be used in connection with the angularity chart, it is necessary that the number of divisions for each stroke correspond. In determining the amount of time to be allowed for the periods of rest and motion, no two designers hold the same opinion. This explains why we see some gas engines with easy riding, quiet running cams, while others have sharp rising, noisy ones. The proportions here described, however, have given excellent results in actual use.

By referring to Fig. 21 it is seen that the piston moves slower at the outer end of the stroke than it does at the start. For this reason, more time can be allowed for the opening of the exhaust valve than can be spared for closing it. In Fig. 22 the exhaust valve starts to rise at division 13 of the out stroke and is wide open on division 14 of the back stroke, remaining open until division 3 of the back stroke is reached. As the piston moves more rapidly at the completion of the exhaust stroke than it does at the completion of the impulse stroke, the cam must fall more rapidly than it rises. In Fig. 22 the exhaust valve is fully closed when the piston is at the end of the exhaust stroke. The inlet valve starts to open while the piston is at the beginning of the intake stroke,

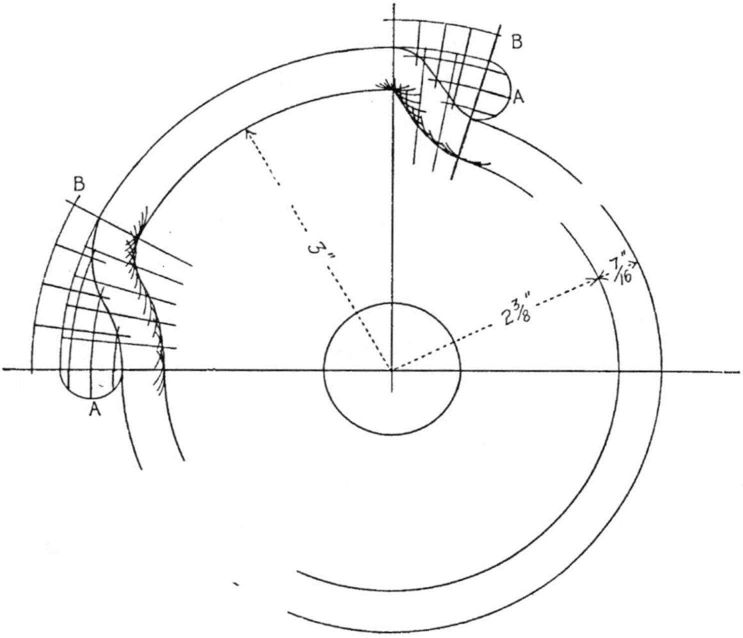


FIG. 23

and is wide open on division 5. It starts to close on division 13, and is fully closed at the completion of the intake stroke.

In Fig. 23 is shown the layout for the exhaust cam. Both the rise and fall are true harmonic curves, developed from the semi-circles *"A A"* and the angle of motion lines *"B B."* In laying out the harmonic curves, the semi-circles (of diameter representing the rise) and the angle of motion lines (representing the period during which the rise occurs) are divided into the same number of equal parts. Straight lines are drawn from the divisions of the semi-circles to their axes, and the curves representing the travel of the center of the cam roll pass through points where the concentric arcs drawn from the axes and radial lines drawn from the angle of motion line meet.

As before stated, the design of cams for the slow running, horizontal type of gas engine calls for practically the same methods as those used in designing machine cams. Designs for cams for automobile engines have, however, been developed by experimenting with various shaped cams, from which formulas were afterward standardized.

In determining the periods of rest and motion for automobile engine cams, no two engineers agree. It is common practice, however, to have the exhaust valve start to open when the crank pin is approximately 25° from its lowest point on the impulse stroke, the valve not fully closing until the crank pin has travelled 10° on the intake stroke. This is shown in Fig. 21. The inlet valve starts to open as soon as the exhaust valve is fully closed, and closes at the completion of the intake stroke, or sometimes a few degrees afterward. Anyone who has had occasion to examine the valve motions of several leading makes of automobile engines will admit that the timing of the valves varies, sometimes to a considerable extent.

In designing any automobile engine cam, the first step is to decide upon the periods of rest and rise, as these two factors determine the time that can be allowed for rising, remaining open, and closing. As before stated, it is common practice to open the exhaust valve 25° before the completion of the impulse stroke, and to close it 10° after the completion of the exhaust stroke, or 10° on the intake stroke, which is one and the same thing. By referring to Fig. 21 it is seen that the piston has very little movement during the time in which the crank pin travels over the degrees above mentioned; therefore very little, if indeed any, power is wasted.

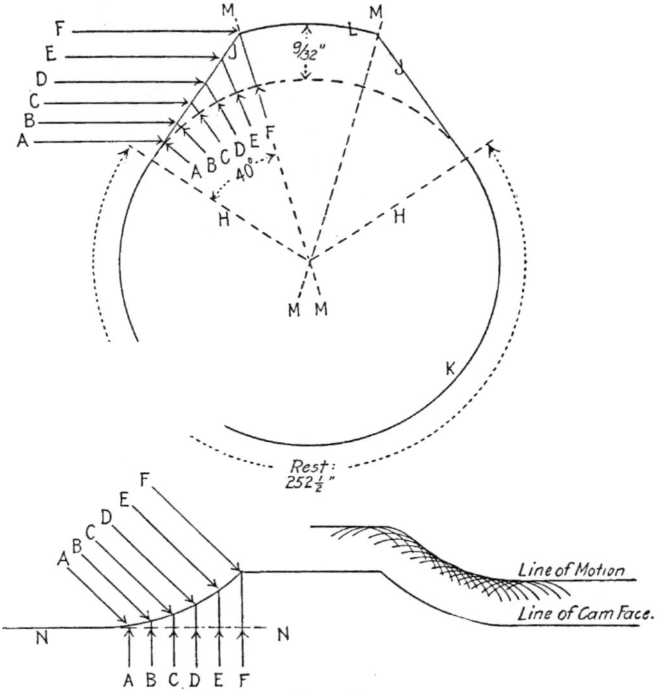


FIG. 24

As an example, let it be assumed that the exhaust valve is to open when the crank pin is within 25° of its lowest point on the impulse stroke, and to close after the pin has traveled 10° on the intake stroke, and that a valve rise of $\frac{9}{32}''$ is necessary. In this case, the valve is closed while the crank shaft travels over 170° of the intake stroke, 180° of the compression stroke, and 155° of the impulse stroke, making 505° in all. As the half time shaft travels only half as fast as the crank shaft, it is evident that the cam should have a rest period of $252\frac{1}{2}$°.

The layout of this cam is illustrated in Fig. 24, the lines "*H H*," defining the period of rest being spaced $252\frac{1}{2}$° apart. The lines "*J J*" are drawn at right angles with the lines "*H H*" and therefore tangent to the circle "*K*." The arc "L," which determines the rise of the cam, is $\frac{9}{32}$" from the circle "*K*." The points where the lines "*M M*" intersect the arc "L" determine the length of time that the valve remains stationary while open, which in this case is $27\frac{1}{2}$°. The same method is employed in laying out the inlet valve cam, the only difference being that the period of rest is longer, while the period of action is shorter.

The cam above described is sometimes called a tangential cam because its sides are tangent with its hub. Although this type of cam is used extensively at the present day, it is not modern by any means, as it was used on the famous De Dion Bouton engines, over 20 years ago, in the days when automobiles were called horseless carriages. By referring to the development of this cam, it is seen that the rise is very easy; in fact, it is not unlike the well known harmonic curve.

In laying out the development of this cam, the part of the circle "*K*" that represents the angle of motion is divided into six equal parts, marked "*A B C D E F*," and radial lines from these divisions carried to the periphery of the cam. (These lines point toward the cam's center, which is necessary to obtain a true development). Divisions the same distance apart are now laid off in the lower diagram on the line "*N N*," or inner line of cam face. The distance between the points "*A*," "*B*," "*C*," "*D*," "*E*," "F," is the same in both drawings. A curve representing the line of cam face is drawn by laying off, in the lower diagram, vertical distances equal to the corresponding radial rises of the cam. The line of motion is parallel with the cam face line and distant from it by the radius of the cam roll. As the

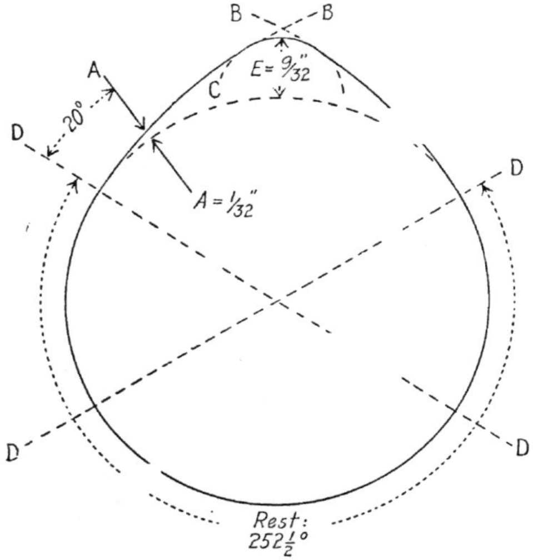


FIG. 25

valve and cam roll move in exact relation with each other, the line of motion thus developed shows the true travel of the valve.

Figs. 25, 26 and 27 illustrate a form of continuous motion cam that is very popular. It is claimed by some engineers that this form of cam is not theoretically as correct as the tangential cam, but as it is seen on many high grade engines, there certainly is some authority for using it. The cam shown in Fig. 25 is laid out by laying off the period of rest in the same manner as described with the tangential cam: that is, the lines *"DD"* are spaced 252½° apart. The arcs *"BB"* are drawn from centers on the lines *"DD,"* the radii of these arcs being determined by the radius of the arc "*C*," which has its center on the circumference of the main circle. The dimension "*E*" is the stroke of the cam, that is $\frac{9}{32}$".

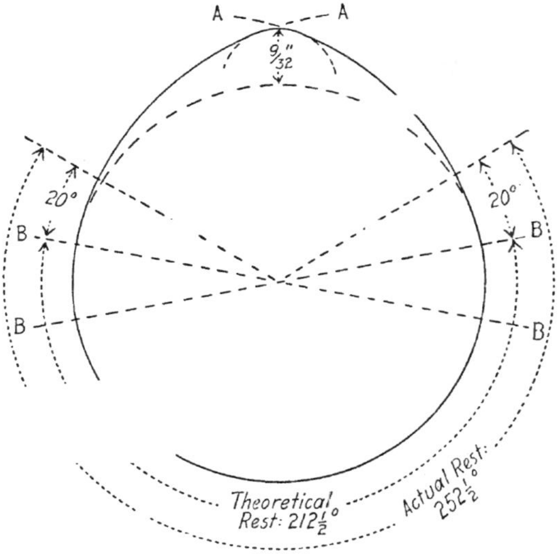


FIG. 26

This is a poor design as the rise is not rapid enough. Notwithstanding that the valve rises the required distance from its seat, it is not fully open long enough to free the cylinder of the burned gas. Again, as it opens slowly, owing to the fact that the arcs "*B B*" do not impart any noticeable lift to the valve when it starts to open, the factor of "wire drawing" is present and will cause excessive wear of both valve and seat. Further, there is always some backlash between the push rod and the valve stem, also between the cam roll and the stud, especially in cases where the engine has been in use for some time without due attention having been paid to adjustments.

The factor of backlash is overcome in the cam shown in Fig. 26. *H*ere the theoretical rest is shortened 40°, the arcs *"A A"* being drawn from points on the lines *"B B."* Even with this change, the factor of excessive wire drawing is not overcome because the valve still opens

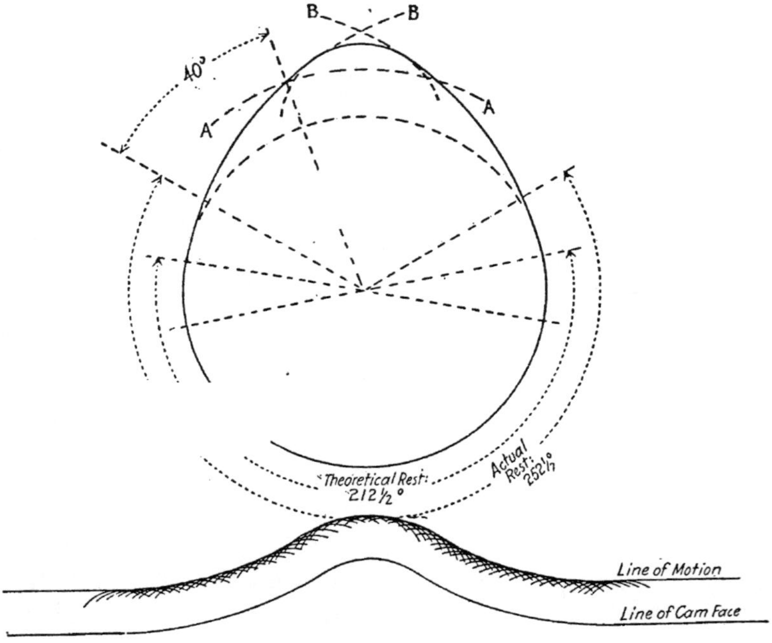


Fig. 27

slowly; in fact it is not open the required distance until the cam is almost at full stroke.

Both objections referred to are overcome in the cam illustrated in Fig. 27. *H*ere the valve is raised to the required distance after the cam has travelled 40° from the actual period of rest. The terms actual and theoretical rest are here used to make allowance for backlash. The distance from the arc *"A"* to the top of the cam is, of course, over-travel. The object of this is twofold: It keeps the valve in continuous motion and tends to eliminate wire drawing, as the arcs *"B B"* rise more abruptly from the cam circle than do the arcs *"A A"* in Fig. 26.

A continuous motion cam is claimed by some engineers to be a better form than the tangential cam, in which the valve starts and stops twice between the periods of rest, whereas in the continuous motion, the valve starts and stops but once. The only disadvantage that the continuous motion cam is said to possess is that the long stroke throws an undue strain on the exhaust valve spring. If the springs are properly designed, however, and are of sufficient length, no serious trouble is noticeable. The cam illustrated in Fig. 27 imparts a very easy motion, as the development shows.

# CHAPTER III

## Cam Followers

Cam followers are generally spoken of as the rolls or fixed parts, as the case may be, that come in direct contact with the cam face, or path. In a broad way, however, cam followers include all the parts utilized to convey the motion from the cam development to the parts that are actuated. Thus rolls, fixed followers, cam bars and cam levers all come under the head of cam followers.

A fixed follower, as shown at "*A*" Fig. 28, is often used in connection with the exhaust cams of motor-cycle, and other small, internal combustion engines. As this follower is directly fastened to the push rod by means of a tapered pin, it would at first appear that unnecessary parts were eliminated. This is not the case, however, for in glancing at "*A*" and then at "*B*," where a roll is used, it is seen that there are the same number of parts in each case.

The fixed follower creates an unnecessary amount of friction on the cam face, while with the roll follower the wear comes on the roll pin. In cases where the roll, pin and push rod are made of sufficient dimensions the roll form of follower is to be preferred. The mushroom form of push rod, shown at "*K*," eliminates some of the friction, as it rotates in its housing. It is generally made in one piece, and has found favor with many designers.

With the wiping cam shown at "*C*," a fixed follower is necessary as a sudden drop is desired. If a roll were used here, a wedging between the roll and the drop on the cam face would result. An eccentric cam, as shown at "*D*," calls for fixed followers in the majority of cases, but, as before stated, it is sometimes used in connection with rolls. In cases where it is necessary to employ

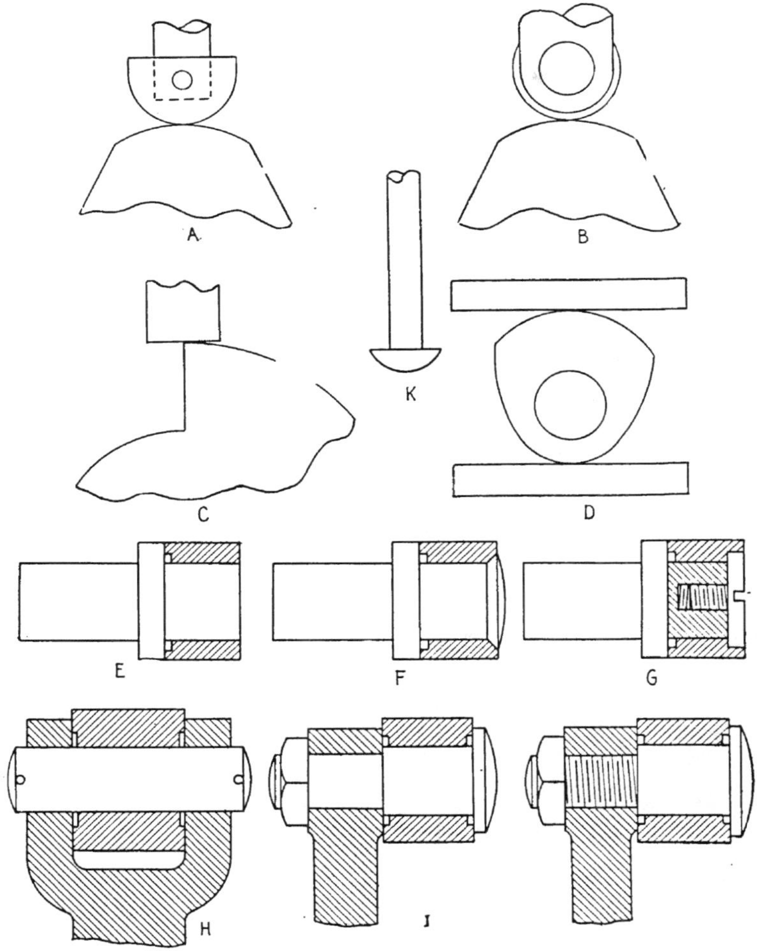


FIG. 28

fixed followers, both the cam and follower, or followers, should be case-hardened, if this is practicable, as friction is thereby greatly reduced.

With the majority of cams, it is the accepted practice to use roll followers. The simplest form of roll and stud is shown at *"E."* As this type of roll and stud is designed for use with a path cam it is not necessary to provide means to prevent the roll from working off the stud. The stud is fastened to the cam lever by means of a drive fit.

The rolls shown at *"F," "G," "H," "I," "J,"* are designed to be used in connection with open cams. In this case it is necessary to provide means for keeping the rolls in position on their studs. A simple means to this end is shown at "F," the end of the stud being headed over into a counter-sunk depression at the end of the roll. This form of construction is simple and efficient and is often seen in shoe and sewing machine construction.

The form shown at *"G,"* wherein the roll is held in position by means of a screw, is sometimes used. This form is more expensive to construct than the one just described, and aside from the fact that the roll can readily be removed if occasion should require, it possesses no advantage. After a machine is once assembled it is seldom necessary to remove the rolls from their studs as the wear on each is about equal, thus necessitating a new stud as well as a new roll, when repairs are to be made. For this reason, the form shown at *"G"* is seldom preferred to the one illustrated at *"F."*

The form shown at *"H"* is often used in high grade machine construction. As the roll pin is supported on both sides at equal distances from the point of applied power, unnecessary side strain is eliminated. This form is very simple to construct as the stud is a straight pin, held in position by means of a drive fit and two spring cotters.

The form shown at "*I*" is often seen but it can not be claimed that this design possesses particular merit, and it is expensive to construct. In some cases, however, where it is necessary to remove the stud from the lever, for repairs or other purposes, there is authority for using this design.

The form shown at "*J*," wherein the stud is screwed into the lever, is expensive to construct, one reason for this being that the lever must be tapped accurately, otherwise the roll will not have a true bearing on the cam face. As the lock nut prevents the stud from working loose under severe strains, these threaded studs are often used. Unless limited space has to be taken into consideration, none of the various forms used in connection with open cams are as practicable as the one illustrated at "*H*."

Cam rolls are sometimes made of tool steel, hardened and drawn to a medium straw color. It is more economical, however, to make them of low carbon steel, carbonized to a sufficient depth to insure reasonable wear. They can be formed on the screw machine by methods known to any mechanic. One point should be given consideration; that is, the ends should be counter-bored. This precaution, simple as it may seem, is the means of preventing the rolls from sticking to their studs, thereby wearing flat. With the design shown at "*E*," it is necessary to counter-bore one end only. With the other designs shown it is necessary to machine both ends, which calls for an extra setting in the screw machine. The advantage gained, however, will offset the extra production cost.

There are three methods in common use for finishing the holes in cam rolls, *viz.:* lapping, grinding and grinding followed by lapping to remove the wheel marks. The amount to be left for lapping depends on the size of the hole. For holes $\frac{1}{4}''$ and under, 0.001″ is sufficient, that is if the holes have been properly reamed. For holes $\frac{1}{4}''$ to $\frac{1}{2}''$ 0.002″ will be sufficient. With holes $\frac{1}{2}''$ and over it is best to finish by other methods.

Three forms of laps used for finishing cam rolls are shown in Fig. 29. *"A"* illustrates a simple form of lap made of copper or brass. This lap is split and provided with a wooden wedge for expanding, to take up wear. This is not the best form of lap to use, and it will not give the most efficient results, even in the hands of a skilled workman.

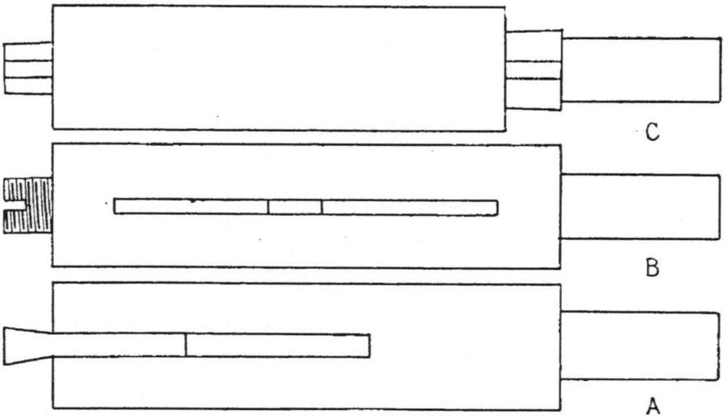


Fig. 29

Another form of lap is shown at *"B."* This lap is split by means of a milling saw, and it is provided with a taper ended screw for expanding. A better form is shown at *"C."* This lap is made of a composition of three parts lead to one part tin. It is split for its whole length and is mounted on a tapered spindle. The wear is taken up by driving the lap up the spindle, thus expanding it. The spindle has a half-round groove milled on it to prevent the lap from turning.

Emery is the abrasive commonly used for lapping cam rolls, as it is cheap and easily procured. Some of the more recent, electrically-made alumina abrasives will give better results, as they cut faster, owing to the fact that oxide of iron, one of the component parts of emery, is practically eliminated.

Care should be exercised in lapping the rolls, otherwise they will be bell-mouthed to such an extent as to lower their wearing qualities. The rolls should be washed in gasoline before testing with a size plug; otherwise a few grains of the abrasive would soon wear the plug under-sized. In cases where it is not necessary that the rolls be interchangeable it is only necessary to lap them enough to make the holes smooth. Rolls of the type shown at "F," Fig. 28, are seldom made interchangeable.

The holes can be finished very rapidly and accurately, by grinding either on a universal grinder or on a regular internal grinder. The latter is to be preferred, as it is a machine designed to do one class of work in the most efficient manner. It is common practice to hold the rolls in an ordinary three jaw chuck. While this method is simple, the production cost is unduly high as it is generally necessary to true up each roll to a certain extent, otherwise they will not finish out.

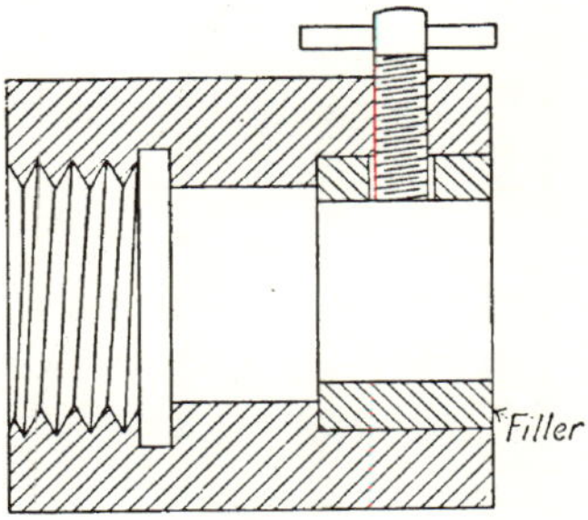


FIG. 30

The fixture shown in Fig. 30, which screws on the headstock-spindle, will give better results as the rolls can be located instantly. Several fillers can be provided to accommodate different diameters of rolls, and the time saved in locating the work soon offsets the cost of providing the fixture.

The allowance to be left for grinding depends on the finish left in the holes as they come from the screw machine and the accuracy of the outside diameter. If some of the rolls are a little smaller than the internal diameter of the filler in which they are held in the grinding operation, this will, of course, throw the hole a little out of center. Under ordinary conditions, however, with holes ½″ and under, an allowance of 0.003″ to 0.005″ should be left for grinding. With holes from ½″ to 1″, 0.005″ to 0.010″ will be sufficient. With larger holes, the proportion can be increased with the diameter. Wheels made of pure corundum were at one time considered the best for internal grinding of cam rolls. Under present-day conditions, however, several of the artificial alumina abrasives have been known to give excellent results. The grit should be 60 to 70 and the grade medium-soft to soft. It is impossible to give exact rules for the selection of wheels for this work as local conditions have to be considered in nearly every case.

If we examine with a microscope, any surface left by a grinding wheel, no matter how fine, it is seen that the surface presents innumerable scores that are left by the grains of the abrasive. For this reason, many manufacturers of high-grade machinery finish cam-roll holes by lapping after grinding. The lap shown at "B," Fig. 29, when used with abrasive in flour grit, will give good results on this work. As it requires less than 0.0005″ to lap out the marks left by a wheel in 60 grit, this final lapping consumes but little time, and the rolls thus finished are practically insured against roughing up on their studs. Therefore, this practice is to be recommended.

The outside of the rolls should always be finished by grinding. The wheel used for this purpose should be of the same material mentioned for internal grinding, but it should not, as a general rule, be finer than 60 grit. The grade used in this case is generally medium. The work can be done on either a plain or universal grinder. The rolls can be held on arbors as shown in Fig. 31.

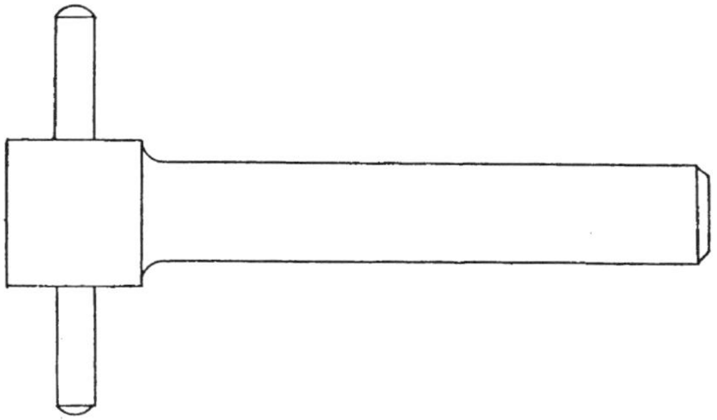

FIG. 31

These arbors should have a slight taper for gripping the work, and are provided with a driving pin that takes the place of the usual dog.

The most efficient method for grinding cam rolls is to use a wheel with a face slightly wider than the length of the rolls to be ground. The wheel is trued carefully by means of a diamond, fed past it automatically. In grinding, the traverse feed is discarded entirely; the work being fed directly to the wheel. It is, of course, necessary to true the wheel once in a while; but if it is properly trued, one hundred or more rolls can be ground after each dressing.

This method of grinding is still in its infancy owing to the fact that until recently it was almost impossible to secure satisfactory wheels to use in the manner described. At the present time, owing to the fact that keen competition has compelled grinding wheel manufacturers to furnish satisfactory wheels for difficult grinding jobs, the above method of grinding is coming into common use, and the results are satisfactory.

Roll studs can be finished in two ways: by grinding, or filing in the speed lathe. Grinding is to be reccommended in cases where interchangable parts are required, or in cases where it is necessary to reduce production costs to a minimum. At the same time, we must not loose sight of the fact that studs and rolls can be fitted in a very satisfactory manner by filing, if the work-

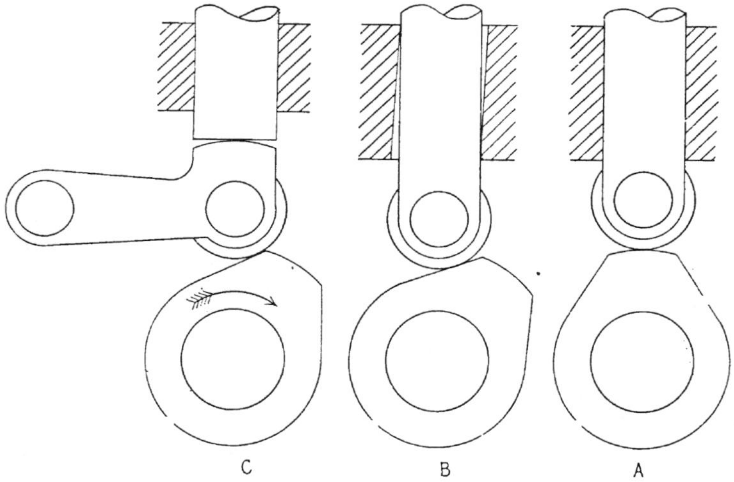


FIG. 32

man possesses the necessary skill. If the studs are to be ground, the method described for grinding the rolls can be used to advantage. It is, of course, necessary to nick the work at the shoulders to provide a space for the corner of the grinding wheel, otherwise a fillet would result. About 0.005 should be left for grinding on all sizes up to 1″. For larger diameters, a more liberal allowance should be made. If the studs are to be finished by filing, care should be exercised to see that the finishing mill, used on the screw machine for forming the studs, leaves a smooth surface, about 0.002″ oversize. A No. 4 safe-edge Swiss, or Swiss pattern, file should be used as the workman should take enough time in fitting to turn out reasonably accurate work.

The simplest method of converting cam motion into reciprocating motion is by means of a rod or bar whose center-line is in line with the center of the cam shaft. An illustration of this is shown at "*A*," Fig. 32, which

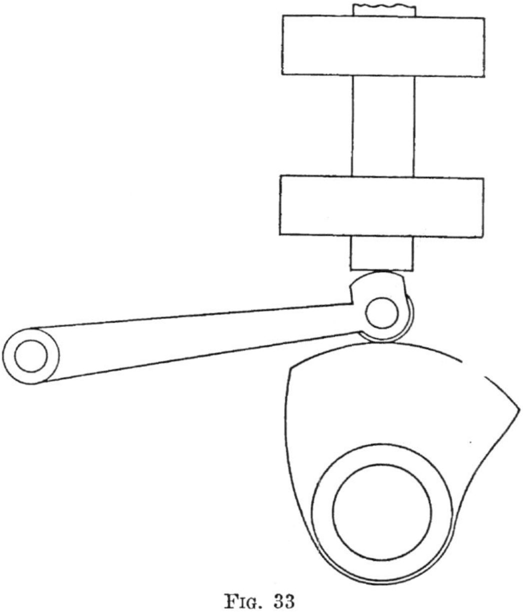

Fig. 33

represents a cam and push rod used on internal combustion engines. This motion is apt to create unnecessary friction and is therefore short-lived, owing to the fact that the cam throws a side strain on the push rod, wearing its housing as shown at "*B*." This causes pounding and also alters the timing, making the valve late in rising. This disadvantage can be overcome by interposing a swinging lever as shown at "*C*." The cam should rotate in the direction shown by the arrow, thus relieving the roll, stud, and lever fulcrum of undue strain. If the cam were revolved in the opposite direction, a cramping effect would result.

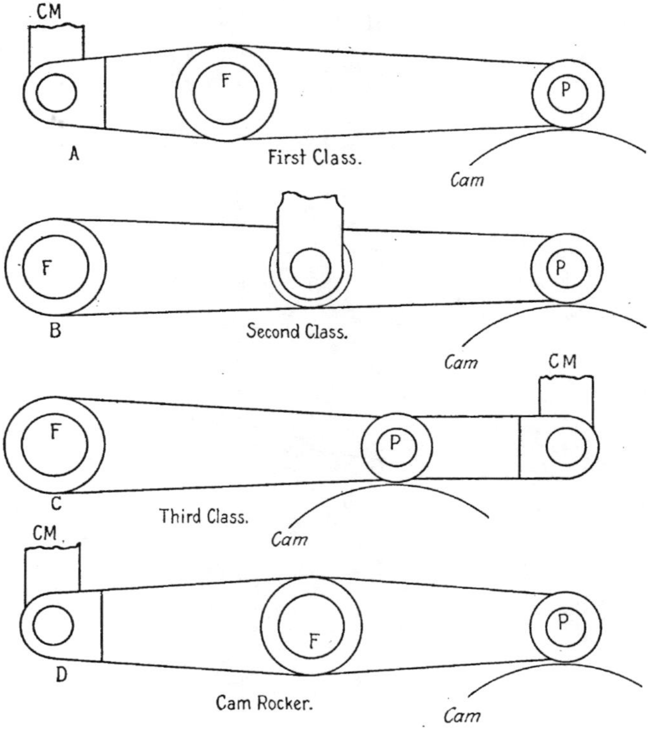


FIG. 34

The distance from the center of the lever to the center of the cam shaft should equal the small radius of the cam, plus one half the stroke. This lever can be made of steel, malleable iron, or of a steel casting. The surface that comes in contact with the push rod should be case-hardened to eliminate friction as much as possible. This same principle can be employed to advantage on machine cams of comparatively short strokes. An illustration of this is shown in Fig. 33. Without the interposed lever, an undue strain would be thrown on the roll, stud, and the brackets that confine the cam rod.

The most common method of converting cam motion into reciprocating motion is by means of levers of the first, second, and third classes. An illustration of these different levers as applied to cams is shown in Fig. 34. In this illustration, the letters *"CM"* stand for "converted motion," taking the place of the letter *"W,"* which stands for weight in ordinary problems of leverage. The rocker shown at *"D"* is often employed as a cam follower. Many designers speak of this rocker as a lever but it can not strictly be called such as no force is gained. It is a rocker for conveying and reversing motion and nothing more.

In machine design, a mistake is often made in making cam followers of any convenient shape that will conform to the general design of the machine. The cam followers should receive first consideration. Where cam motions are necessary, the motion should be conveyed to the parts to be actuated with as little friction as possible.

A common error is shown at *"A,"* Fig. 35. This is not a lever in the way it is used because the fulcrum is too near the point where the power is applied. In a machine running even at a moderate speed an undue strain would be thrown on the cam and its roll and stud, owing to the fact that the distance from the fulcrum to the point of converted motion is greater than the distance from the fulcrum to the point where the power is applied. The object of using this motion is to obtain a greater travel than is given by the cam movement.

Where it is necessary to do this, it is considered better practice to use a third-class lever.

A better form is shown at "*B*." This is a rocker wherein the strain is evenly distributed, as the fulcrum is midway between the point of applied power and con-

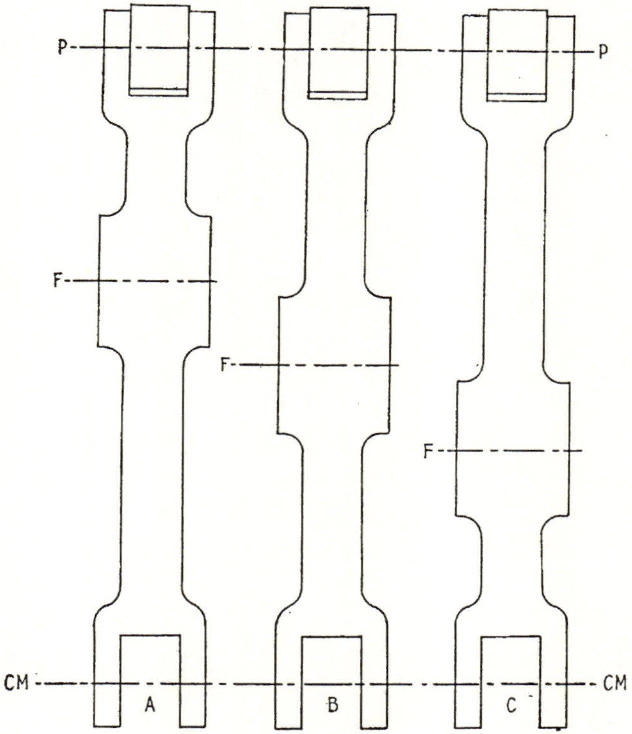


FIG. 35

verted motion. A still better form is shown at "*C*," which is a lever of the first class. This design calls for a greater cam travel to impart the desired motion, but as friction on the cam path is thereby reduced, many designers consider this form the most practicable, and use it when possible. Generally speaking, cam levers can be of the class best suited to the general design of the machine, but the third class lever should be used as little as possible, and the error shown at "*A*," Fig. 35 should never be committed.

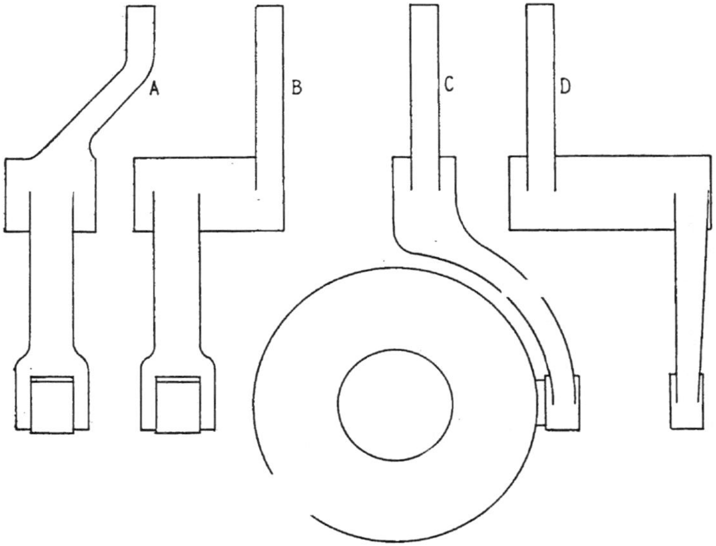


Fig. 36

In cases where it is necessary to offset the line of motion, a lever, as shown at "*A*," Fig. 36, is sometimes used. This is another example of poor design. A better way is shown at "*B*," in which the motion is conveyed in straight lines, which method eliminates cramping. In this case, the stud that forms the lever fulcrum should be of sufficient dimensions to convert the motion without undue side strain.

Another error sometimes seen is shown at "*C*." Here the arm running from the fulcrum to the point of applied power is offset, thus throwing an undue side strain on the roll and cam path. Better results, in this case, would be obtained by using the design shown at *"D,"* especially in cases where it is possible to support the fulcrum stud at both ends.

The best general rule to follow in designing cam followers is to take plenty of time at the drawing board, working along known mechanical principles, with good, every-day common sense. A few extra hours, or even days, spent at the drawing board are often the means of saving thousands of dollars, as expensive delays, caused by the finished machines not coming up to expectations, are thus eliminated.

There are no absolute rules for determining the proportions of cam followers. An engineer can figure to a nicety the velocity of the moving parts and the strength of materials entering into construction, and, by including a factor of safety, he can arrive at approximately definite results. If the results do not come up to his expectations, he increases the factor of safety and redesigns the parts that have proved to be of insufficient strength.

Close-grained, gray iron is an ideal material for cam levers. Good results can also be obtained by using malleable iron. This material, however, possesses one disadvantage in that it cannot always be obtained promptly. Steel castings make excellent cam levers if they are free from blow holes.

# CHAPTER IV

## Master Cams

The words master cam and leader are sometimes misunderstood as both are often used to designate the same thing, that is, a device used as a pattern in cutting cams. According to some authorities there is a distinction, a master cam being made by hand by a skilled workman, while the leader is cut on the cam cutter, using the master cam as a guide.

Sheet brass is the material commonly used for making master cams as it is easily worked, as regards scribing the lines and filing into shape. The lines can be seen more readily if the metal is dipped in any of the well-known solutions for blackening brass. In Fig. 37 is shown the movement illustrated in Fig. 6 laid out on sheet metal $\frac{1}{8}''$ thick to form a master cam from which a leader for a drum cam is to be cut. The movement is as follows: First a rest of $\frac{1}{16}$ rev., then a rise of $\frac{15}{16}''$ in $\frac{3}{32}$ rev., next a rest of $\frac{1}{4}$ rev., after this a rise of $\frac{1}{4}''$ in $\frac{1}{8}$ rev., followed by a rest of $\frac{1}{32}$ rev., and a drop of $\frac{3}{16}''$ in $\frac{3}{8}$ rev. Next comes another rest of $\frac{1}{16}$ rev., which completes the movement.

The master cam blank, after being cut to the correct length to wrap around the drum that it is to be fastened on, is divided into 32, or any other number of equal parts. In this instance 32 divisions are used. The bottom should be straight and the ends square. The skeleton outline of motion is then drawn, and several screw holes, for fastening the master cam in place on its drum, are drilled.

In laying out the harmonic curve for the first arise, a semi-circle is drawn, as shown, lines from the divisions

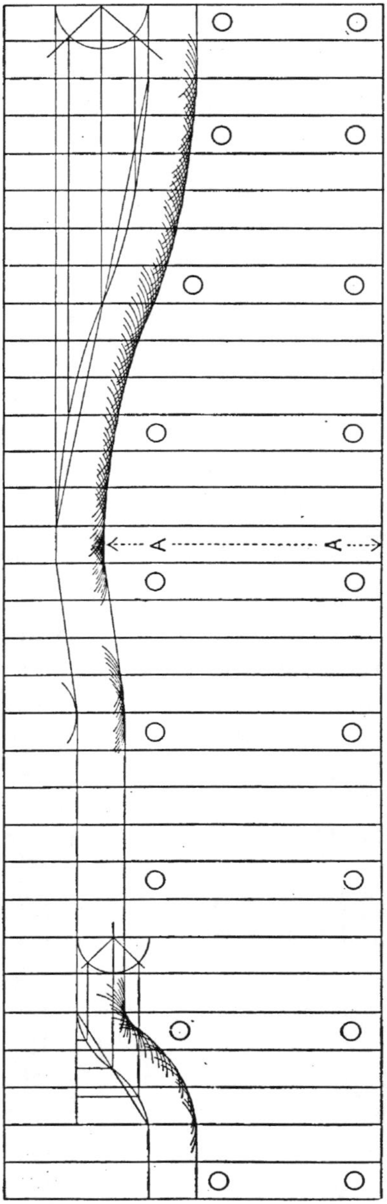


FIG. 37

of which are connected with lines drawn from the motion line. The curve is drawn through the intersections of these lines. For clearness of illustration, only four divisions of the semi-circle are used. In actual practice, however, it is better to use eight or ten. This will insure a correct curve. As the next rise is very slight in comparison to its length, a uniform motion will be satisfactory, care being taken to draw an arc at the start of the rise to avoid cutting back. The radius of this arc should at least equal the radius of the roll that is to be used with the master cam but it is better to make a more liberal allowance when possible.

The last drop is laid out in a harmonic curve by the same method used with the first rise. The next step is to draw numerous arcs from the skeleton line at all divisions that are laid out in curves. The radii of these arcs should be exactly the same as the radius of the roll that is to be used in connection with the finished master cam. This point, which is sometimes overlooked by careless designers, is of the utmost importance, as the skeleton outline represents the travel of the center of the roll. At the periphery the motion is wholly different, owing to rolling motions. Thus, these arcs are necessary to preserve the true motion as decided upon at first. It is also of equal importance that the roll which follows over the master cam is of the correct diameter.

The superfluous metal can now be cut away by any convenient method that will not spring the master out of shape. A sufficient amount should be left for finishing by filing. This final finishing by hand calls for good eyesight, and plenty of time should be taken that a creditable job will result. The master should not be undercut at any point and its finished development should be filed square.

Two castings should be provided; one to fasten the master on, from which the other, the leader, is cut. The casting for the leader should be somewhat smaller in diameter than the one for the master. The height, or thickness of the casting for the master should be about $\frac{1}{4}''$ less than the dimension marked "*A A*," Fig. 37. The outline should be cast on the master blank and leader, as this will save time in making them. These pieces are chucked out and turned up in the regular way, care being taken to see that the circumference of the master blank will just permit the master meeting at the ends when wrapped in place.

It is not as difficult as would be imagined to wrap the master around its casting if the work is done as follows:

First the holes in the center of the master should be transferred to the casting. These are drilled and tapped and the master fastened at this point with screws. Now start to bend the master around the casting, transferring a line of holes at the right, then at the left, and so on

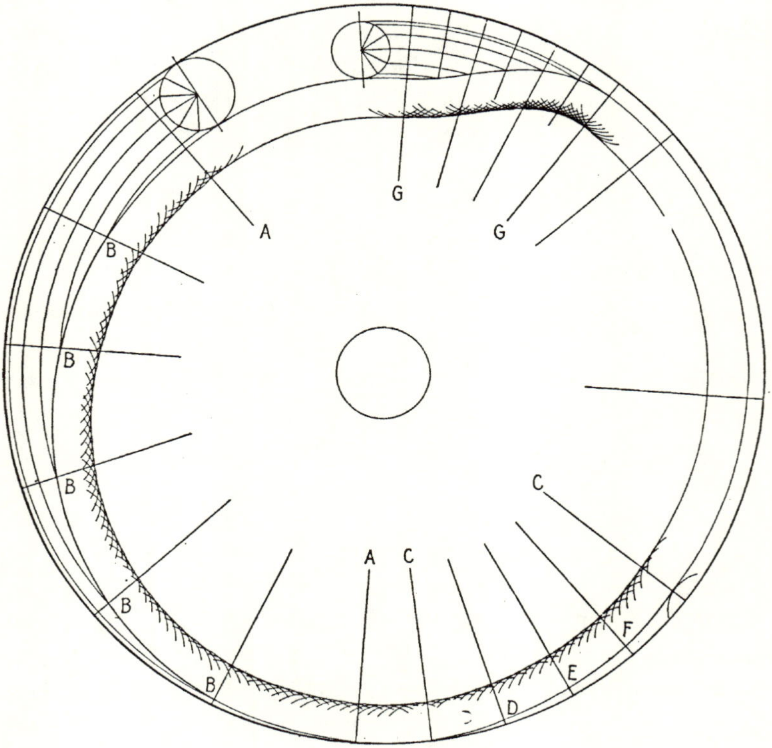


FIG. 38

until all the holes are transferred and the master fastened at all places. The master cam is now ready for cutting the leader.

In Fig. 38 is illustrated the motion previously described, laid out as a master for a disk cam. The process is much the same except that all rises and rests are laid out in curves which have to be drawn correctly to impart the desired motion. The diameter of this master cam should, in all cases, be greater than the diameter of the leader which will be cut from it. A good rule to follow is to make the master cam ¼ larger in proportion than the leader, while the leader should be ¼ larger than the cams. This practice tends to eliminate the factor of chattering, as the angle of the rises are less abrupt as the diameter increases. Thus, the larger in proportion the leader is, the easier it will impart the desired motion in the cam cutting operation.

In making this master, a square piece of sheet brass, of dimensions somewhat greater than the size desired in the finished master, is selected. In the center of this piece, a hole of the correct size to fit the spindle of the cam cutter head is drilled and reamed, after which the piece is roughly cut in the form of a circle. A plug should be provided to fit the center hole; this plug having a center punch mark in its exact center. This mark is used as a guide in scribing the dividing lines. The master is now divided into the number of parts required to lay out the desired motion. In this case 32 divisions are used. For the sake of clearness, however, they are not all shown in the illustration.

As a disk leader is very deceptive, extreme care has to be used in laying out harmonic curves. At the first glance the curve between the lines "*A A*" has the appearance of a uniform motion, as the reverse curve, seen in this same motion on the master for the drum cam, is lacking. The curve is harmonic, however, as it is drawn through points where lines from the divisions of the developing arc and the equal divisions of the motion curve meet at "*B B B B* B."

The uniform rise between the lines "*C C*" is laid out

by drawing a regular curve through the points "*D E* F." As this is a uniform rise of $\frac{1}{4}''$ in 4 divisions of the whole circumference, the distance from "*D*" to the outside circle is $\frac{1}{4}$ of the desired rise, or $\frac{1}{16}''$. From the point "*E*" to the outside circle is $\frac{1}{2}$ the rise, and from point "F" $\frac{3}{4}$ of the whole rise. This method is followed in laying out any uniform motion.

The harmonic curve between the points "*G G*" is now laid out by the methods previously described. As the illustration shows, the curve passes through points where lines carried from the motion circle and radii from the development arc meet.

The radii to represent the travel of the periphery of the roll are now scribed at all motion points after which the master is finished by filing into shape. Care should be taken to mark masters and leaders for disk cams to show which is the outside; otherwise a careless operator is liable to cut several cams backwards through putting the leader in position wrong side out, or backwards. The safest way to guard against this is to mark the correct side: "This side next to shoulder." This infers that the side so marked is to go next to the shoulder of the cam cutting head spindle. If one side is simply marked "outside" no definite directions are conveyed, as the outside could be either side, according to what is considered the front and back of a cam cutter. The master and leader should also be plainly marked with the sized roll that is to be used in connection with them.

Fig. 39 illustrates the layout for a wiping cam. "*A*" shows a uniform rise, while "*B*" illustrates a harmonic curve. The layout for the uniform rise is simple; all that is necessary is to divide the circle into a convenient number of equal parts, 8 are used in this case. On the first division $\frac{1}{8}$ of the desired stroke is laid out, $\frac{1}{4}$ on the

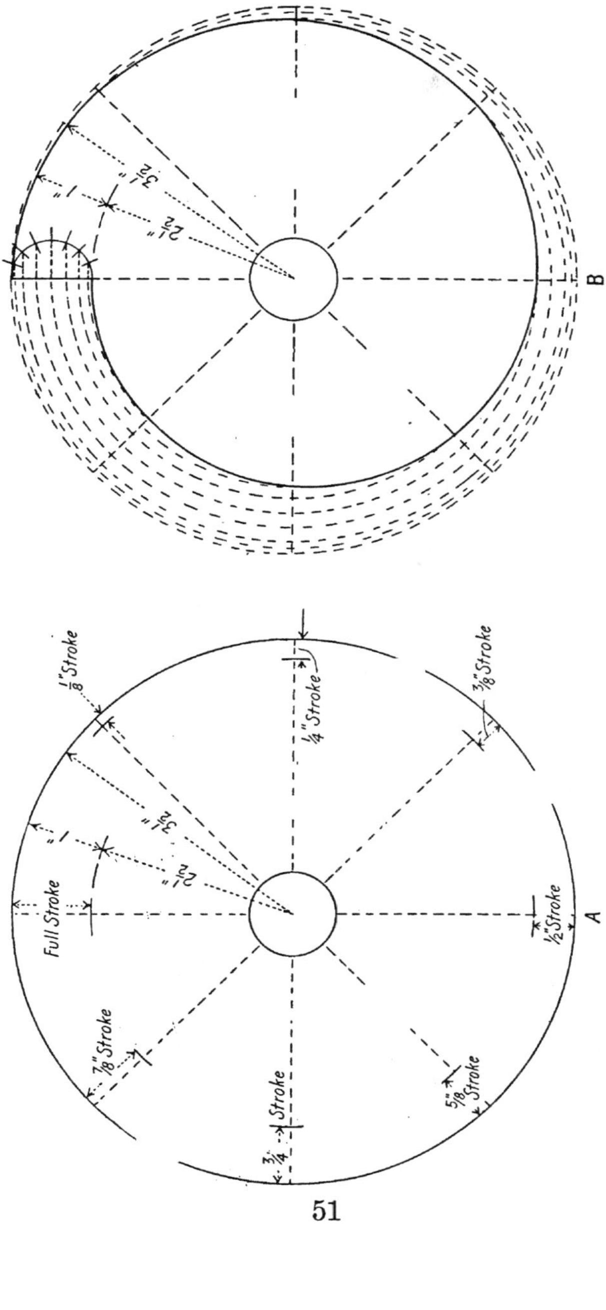
B
A
Full Stroke
1/8" Stroke
1/4" Stroke
3/8" Stroke
1/2" Stroke
5/8" Stroke
3/4" Stroke
7/8" Stroke
1"
2 1/2"
3 1/2"

FIG. 39

next division and so on. A regular curve drawn through these points gives the line of motion.

The harmonic curve shown at "*B*" is laid out according to the methods described with the master cam shown in Fig. 38. The harmonic curve gives a very easy motion to a wiping cam and, therefore, it should be used, unless

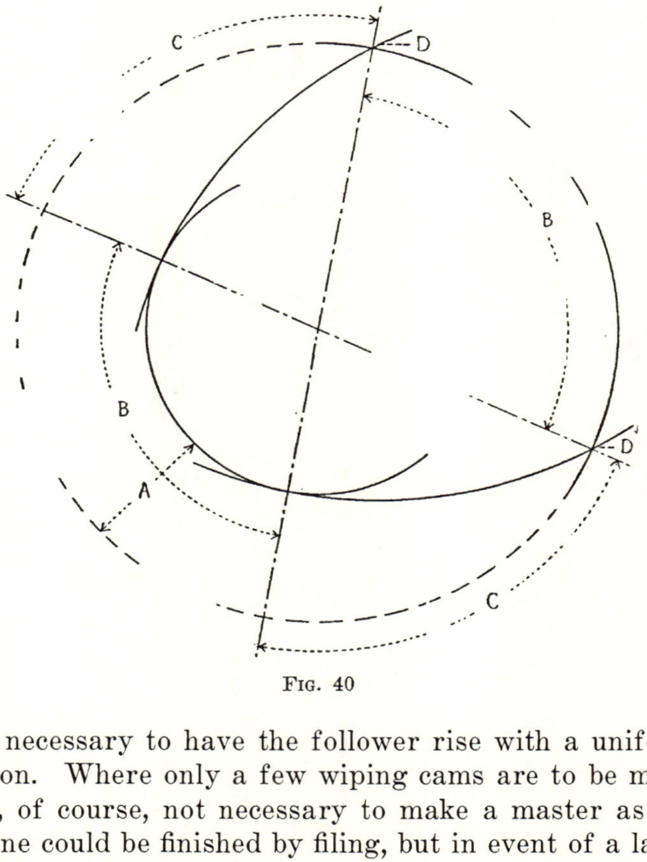


FIG. 40

it is necessary to have the follower rise with a uniform motion. Where only a few wiping cams are to be made it is, of course, not necessary to make a master as the outline could be finished by filing, but in event of a large number being required, a master should be made in the regular way as previously described, and a leader cut from it.

The layout for an eccentric cam is shown in Fig. 40, wherein the line *"A"* equals the stroke, the arcs *"B B"* the periods of rest, and the divisions *"C C"* the periods of action. This type of cam is readily laid out as it con-

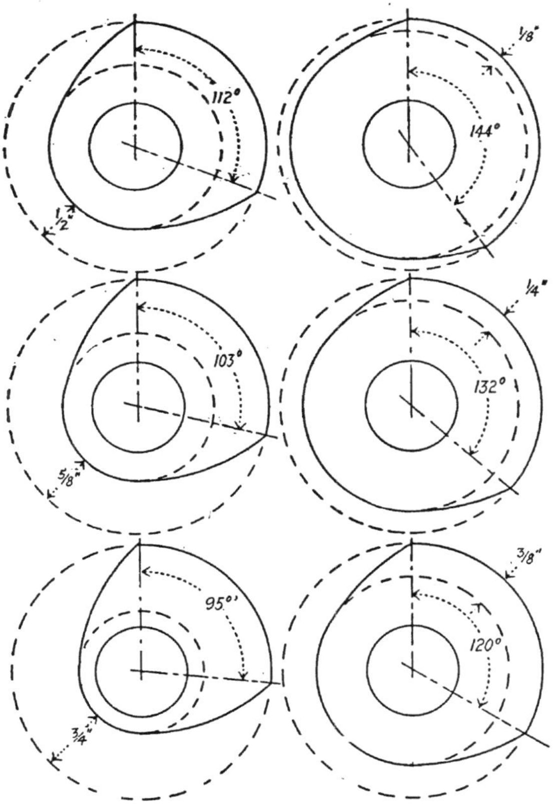


FIG. 41

sists of four arcs only. The arcs representing the periods of rest are drawn from the center, while the radii that represent the periods of action are drawn from the points *"D D"* where the periods of rest end. The period of rest in any eccentric cam is determined by its stroke.

This is illustrated in Fig. 41, where six eccentric cams of the same outside diameter are shown, the strokes ranging from $\frac{1}{8}''$ to $\frac{3}{4}''$. As the illustration shows, the period of rest is shortened as the stroke is increased.

In making masters for eccentric cams by the method followed with ordinary machine cams, the process is long and the results somewhat uncertain. In this case it would be necessary to develop the line of motion by the method shown in Fig. 24, laying out the master from the outline thus obtained. More satisfactory results are obtained by making the master of the size and shape desired in the finished cam, as the method used in developing the leader from the small master cam is simple and the results satisfactory. This method can also be used to advantage in making leaders for cutting automobile engine cams.

Masters for eccentric cams can be partly finished on the milling machine, equipped with a dividing head, but it is generally necessary to finish the arcs between the

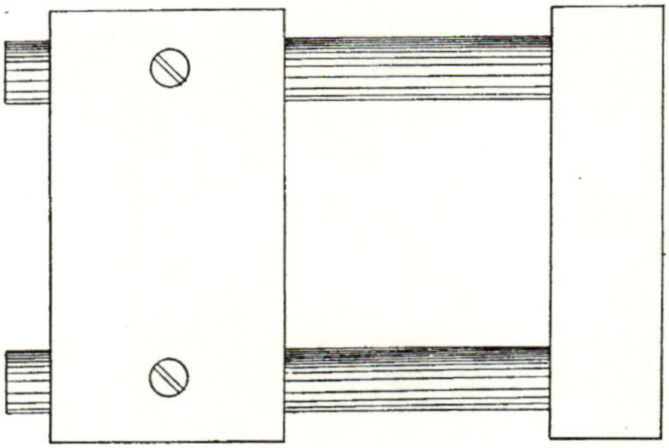

FIG. 42

divisions *"C C"* in Fig. 40 by hand. A gauge as shown in Fig. 42 can be used to advantage in this operation, as it is necessary for the cam to be of the same diameter at all points.

Masters for tangential gas engine cams, which consist of arcs and straight lines only, are readily made on a vertical milling machine equipped with a revolving platen, or by means of a dividing head used on a plain or universal milling machine. With cams of the type shown

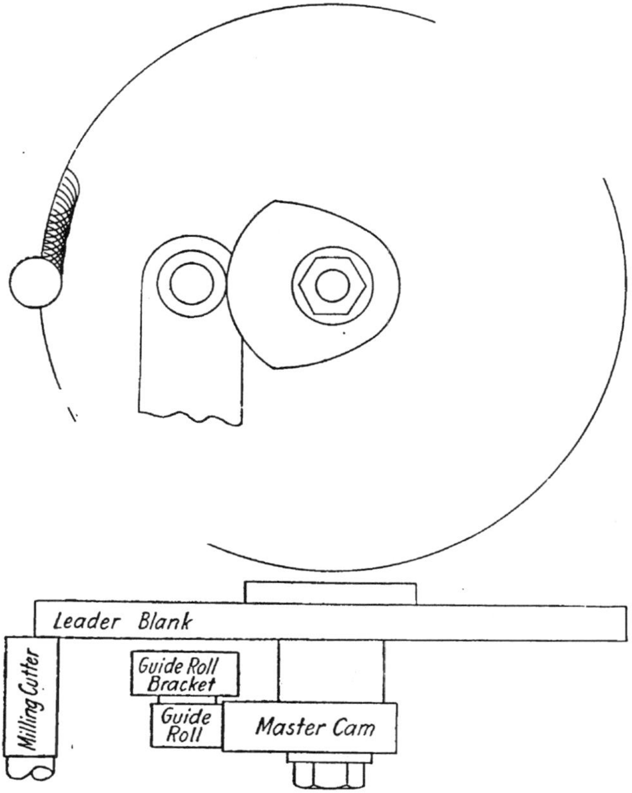


FIG. 43

in Fig. 27 it is necessary to work out the greater part of the outline by hand, which calls for the services of a skilled workman.

In making a leader for cutting eccentric or automobile

engine cams the master is placed on the cam cutting head spindle in the place ordinarily occupied by the cam, while the leader blank is placed in the usual position. The roll bracket is adjusted to bring the guide roll to bear on the master. The guide roll should be of the same diameter as the cutter that is to be used in cutting the cams; as a slight variation here will impart an incorrect outline to the leader. The milling cutter, which must be of the same diameter as the guide roll that is to bear on the finished leader, is placed in the usual position and its end brought against the leader blank. This setup is illustrated in Fig. 43. The cam cutting head is now turned one complete revolution, causing the end of the cutter to mark the desired outline on the leader blank. The leader blank is now taken off, and the superflous metal removed by drilling and filing. The piece is now placed in position again and its periphery carefully milled, the movement being guided by the roll bearing on the master cam. This is considered the most simple and rapid method known for making leaders from small masters and the results are always satisfactory if a reasonable amount of care is used.

From the above description it is seen that with the cams in question we start with a known shape and an approximately unknown motion, from which a leader of unknown shape is developed; the sole object being to generate a leader for imparting the desired shape. This process can be called exactly opposite to the method employed with ordinary machine cams. In that case we start with a known motion from which we make a master of unknown shape, the object being to make a master for imparting the desired motion.

# CHAPTER V

## MACHINE WORK PREPARATORY TO CUTTING AND DIFFERENT TYPES OF CAM CUTTERS

Ordinary machine cams are made of various materials: Cast iron, malleable iron, gun iron, steel castings, drop forgings, and sometimes from steel bar stock, in cases where the cams are small and of the open type. Close-grained cast iron is the material commonly used as it is cheap, easily worked, and possesses good wearing qualities. Malleable iron is sometimes used in cam making, but it can not be claimed that this material possesses any particular advantage. To be sure, malleable iron will withstand a greater bending strain than cast iron, but cams, as a rule, are not subjected to bending strains, and for this reason cast iron is generally preferred.

Gun iron is a good material to use in cam making as it is stronger than cast iron and its cost is not prohibitive. Shoe machinery cams that have to remain accurate under severe strains are often made of this material. Steel castings and drop forgings are excellent materials. for small cams that are to be hardened; the valve motion cams on gas engines being a good example. Steel-bar stock is a good material for small cams as it is readily worked, and it can be case hardened. The methods used for chucking out and turning up cast iron cams for ordinary purposes are simple and need no explanation here as they are known to every mechanic.

Automobile engine cams are generally made of drop forgings, steel castings, or from bar-stock machinery steel. In making cams for experimental purposes, cast iron is sometimes used as it is easily procured and can be readily machined. This material, however, is never used in actual practice. Drop forgings make excellent cams as their use eliminates a certain amount of machine work; but owing to the fact that expensive dies have to be made, and that delays due to material not being delivered on time are often experienced, drop forgings are not universally used. Steel castings are sometimes used for the purpose in question, as they are comparatively cheap and easily procured. Bar machinery steel is a good material to use as it is cheap, easily procured on short notice, and, owing to its soft nature, it can be machined rapidly. As this material can be carbonized

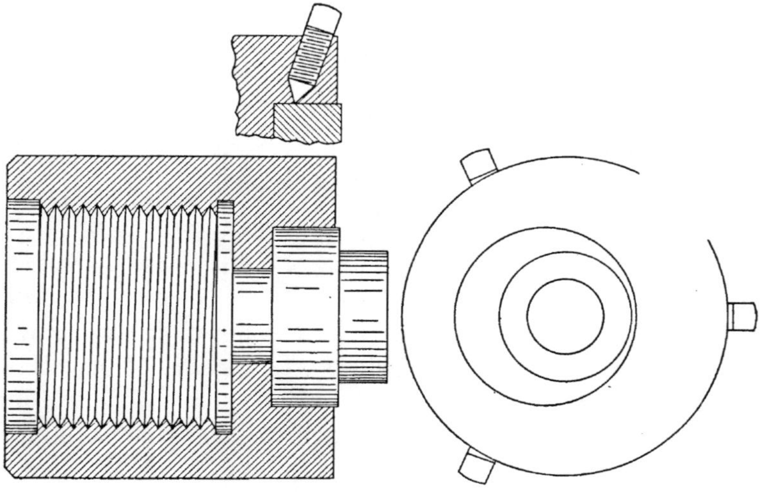

FIG. 44

to any depth it makes long-lived and satisfactory cams.

In making automobile engine cams from drop forgings or steel castings the hole and the plain side can be finished at one setting in the turret lathe. For removing the superflous stock from the other side, and from the hub, good results can be obtained on the cam cutter by substituting a plain disk for the usual leader. As this method calls for little attention, aside from changing the pieces, it is used by many manufacturers.

In making cams from bar stock the material should first be cut to nearly the required thickness; otherwise a long bar of stock would have to be revolved eccentrically in the turret lathe, which is not always practicable. The turret-lathe tools for machining cams are quite simple

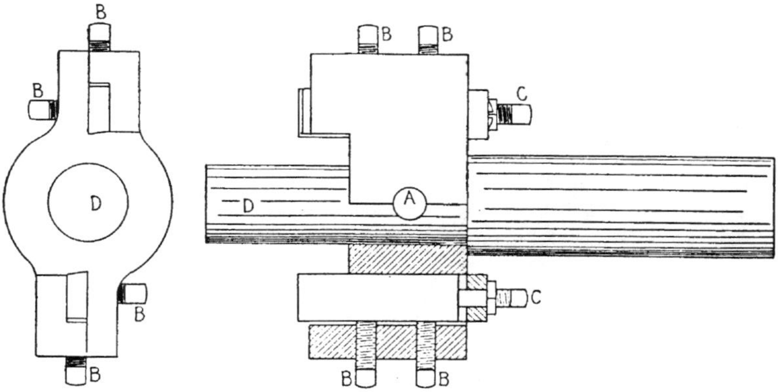


FIG. 45

and can be made by any tool maker on short notice. A chuck for holding the blanks is illustrated in Fig. 44. This fixture is screwed on the turret-lathe spindle. The set screws for gripping the blanks are set at a slight angle as the illustration shows, thus having a tendency to hold the blank against its seat. The blank is spotted, drilled, and reamed in the regular way, leaving a few thousandths for finishing the hole by grinding after the piece is hardened.

For cutting away the superfluous stock from the hub of the cam, the tool shown in Fig. 45 can be used to good advantage. The head, which carries two cutters, is fastened to the shank by means of the taper pin "*A*." The cutters should be made of high-speed steel. They are held in position by the screws "B" and are adjusted by the screws "*C*." When the cutters become dull they can be sharpened without taking them from the head. One end of the shank "*D*" fits the hole previously chucked in the cam and serves as a pilot for steadying the work. This is necessary as the cut is comparatively heavy and off center.

The next point to consider is the manner in which the cam is to be fastened to the cam shaft, as the means of fastening are utilized as a locating point while shaping the face of the cam. Keys, screws, or taper pins, as

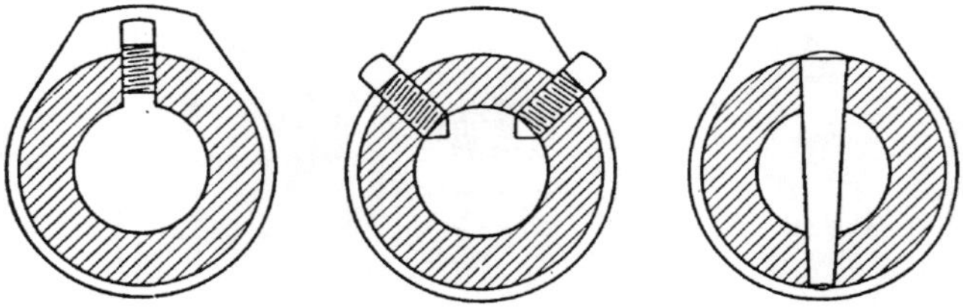

FIG. 46

shown in Fig. 46 are used and all three methods, no doubt, possess advantages. There is one important point that should not be overlooked, *viz.:* all the cams of one kind, that is inlet or exhaust, should be alike as regards the relative position between the means of fastening and the face of the cam. In cases where multiple-cylinder engines are made this makes a reduction in production costs while shaping the face of the cams. Again, the necessity of carrying in stock cams for each cylinder of multiple cylinder engines is eliminated. The differences in position of the cams in relation to each other that are necessary for timing the various valves are fixed by spotting, keyseating, or drilling and taper-reaming the cam shaft, as the case may be, at the correct places to give the proper timing.

Taper pins are considered satisfactory for locating the cams and a simple jig for drilling and reaming the cams is shown in Fg. 47. The cam blank slips over the stud *"A"* and is located by the posts *"B,"* which are machined to fit the outline of the blank. After drilling, the bushing *"C"* is removed and the hole reamed to the

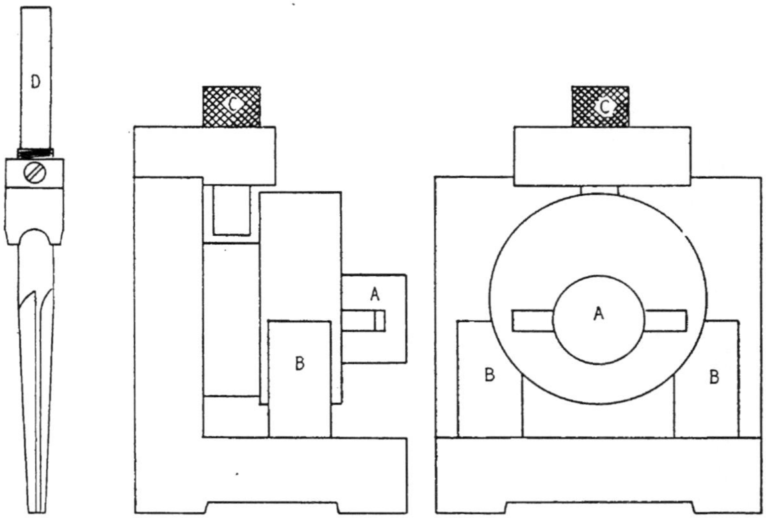


FIG. 47

required depth by the reamer *"D,"* which is provided with a stop. This stop comes in contact with the lining bushing and it is adjustable; this being necessary as it has to be re-set every time the reamer is sharpened.

The next point for consideration is the shaping of the outline or contour of the sam. This can be done on the cam cutter, or on a plain grinding machine equipped with a suitable cam grinding attachment. On this point there is a wide difference of opinion as to what is the best practice. With the modern, substantially built cam cutting machine, the cams can be readily formed from round stock as left by the screw machine. After hardening, the contour can be finished on a disk grinder, thus insuring a square face. Even if very slight irregularities in contour result they will not be of enough importance to effect the value motion materially.

On the other hand, the makers of grinding machines claim that grinding is an economical method to use for roughing cam blanks from round stock and finishing after hardening. Grinding with a disk wheel certainly presents one objection in that the contour of the cam is bound to change as the wheel wears away. When confronted with the problem of installing machinery for making automobile engine cams, it is best to investigate the various methods closely, and to be guided by good common sense.

There are quite a few devices and special machines on the market for cam cutting, and they all work on the same principle: that is, the cam outline, or path, is generated by a follower bearing on the periphery of a leader which is cut from a master cam.

A simple form of cam cutting attachment is illustrated in Fig. 48. The spindle, which is driven by worm gearing, is of ample dimensions and runs in bronze boxes that are adjustable for wear. The cam and the leader are held on an arbor that fits the spindle through the medium of a tapered shank. The head carrying the spindle is bolted to the platen of the attachment and can be set at right angles for cutting drum cams when occasion requires. The roll bracket is adjustable to accommodate leaders of different diameters, and is bolted to the inside of the base, as the illustration shows. For cutting drum cams a different roll bracket is used. The base of the attachment is fastened to the milling machine platen in the regular way, and is provided with gibs for

FIG. 48

compensating for wear. This attachment was originally designed to be fed by hand, but some enterprising mechanic substituted the belt drive shown in Fig. 49. This type of cam cutting device is not modern by any means

FIG. 49

as it has been in use in the eastern states for many years. Why the maker did not think it worth while to attach a

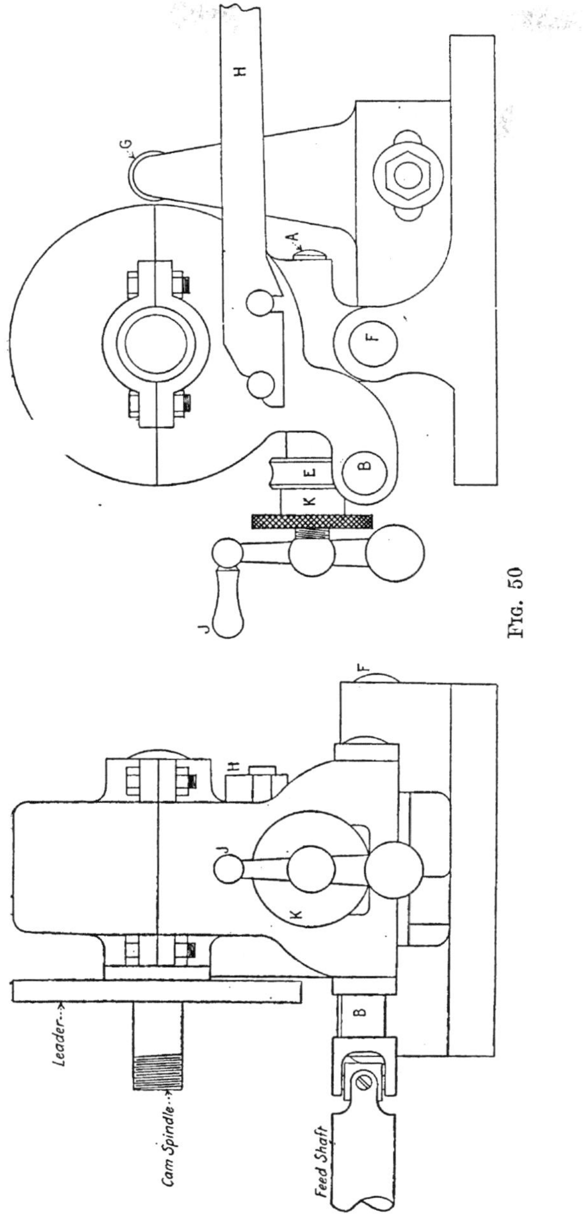


FIG. 50

name plate to this product is a question of conjecture as the attachment is efficient and not high in price.

Another practical cam cutting attachment is illustrated in Fig. 50. This device is of the swing-frame type, the frame being pivoted at its base. It is sometimes claimed that this arrangement will not cut cams that are true to the leader, owing to the swing motion. A little thought will convince us that this assertion is not well grounded, for if the leader is cut on this machine from a master that is correct, any errors due to the swing motion cannot develop. If, however, a leader should be cut on this attachment and afterward used on a cam cutter of the type previously described, a faulty contour in the finished cam curve would result. The use of this device is limited as it will cut face cams only.

The spindle is driven by a train of worm gearing, the main gear being enclosed in the frame casting. The main gear is driven by the worm shaft "*A*," this in turn being driven by the worm shaft "*B*" and worm wear "*E*." The frame is pivoted at "*F*" and the leader is held in contact with the guide roll "*G*" through the medium of the lever "*H*" to which a weight is attached. The handle "*J*" is for hand feeding, the automatic feed being disengaged by the lock nut "*K*." The feed shaft "*B*" is driven by a universal joint shaft actuated by the feed mechanism of the milling machine on which the attachment is used.

In Figs. 51 and 52 is illustrated a modern machine designed and built by the Garvin Machine Company to meet a popular demand for a minimum-sized cam cutting machine. This machine is well adapted for cutting face and drum cams, also for the use of lubricant, in cases where cams are to be cut from steel. The general design follows closely the standard as used on the Garvin

Machine Company's well-known 24″ and 36″ sized profile machines.

The spindle slide is sensitive, running on large steel balls in steel tracks. This feature insures exact duplica-

FIG. 51

tion of the desired contour as the factor of faulty contact, owing to a hard working slide, is eliminated. The drive is from an overhead counter-shaft through the medium of a universal joint shaft.

The rail slide is nicely balanced, being operated by a rotary nut equipped with a ball thrust. This feature gives a rapid motion for the removal of the cutter and guide roll at the completion of the cam cutting operation.

FIG. 52

While changing cams, the spindle slide is clamped, and at the completion of the change the slide returns to a positive micrometer stop. All the gears on this machine are protected by adequate guards, thus conforming to the laws passed in many states regarding the safeguarding of machinery.

The spindle bearings are of the type used on profile machines and are provided with means for compensating for wear. The feed can be disengaged or reversed as occasion requires. The fixtures for cutting face cams can be readily removed and the attachment for drum cams substituted. Fig. 51 shows the machine set for cutting face cams, while Fig. 52 illustrates the drum cam fixture in position.

This machine is compact, occupying less than a square yard of floor space. The design is a radical departure from that commonly seen in cam-cutting practice as the leader and cam are fastened to separate revolving tables. This is, indeed, a decided improvement as, owing to the secure support, the factor of chattering is practically eliminated, thus making it possible to force the cutter to its maximum output.

# CHAPTER VI

## Cam Cutting

The cutters for machining cam contours do not differ in design from ordinary end mills. The teeth may be spiral or straight, the shanks straight or tapered, and the material carbon or high-speed steel. High-speed steel cutters are, of course, the most economical even at their advanced cost as their use reduces production costs to a minimum. Spiral cutters are preferred since their use eliminates chattering. The spiral on a right-hand cutter should be left hand. This feature insures smooth cutting. Where right hand cutters with right hand spirals are used, the tendency is to pull the cutter away from its holder and, if tapered shank cutters are used, broken cutters and spoiled work will often result. Three of the cutters shown in Fig. 53 are right-hand cutters with left-hand spirals. Although a left-hand spiral on a right-hand cutter causes the end teeth to cut with a slight drag, this is of little consequence.

In Fig. 53, "*A*" illustrates an ordinary taper shank cutter. This type possesses one advantage; that is, it can be obtained from any supply house on short notice. Otherwise no particular advantage is gained by its use. A bevel-shank cutter is shown at "*B*." In making these cutters, the shank beyond the bevel is left soft, the thread being finished after the cutter is hardened. As the work on the bevel, thread and teeth is finished on centers, after the cutter is hardened, all parts are concentric. This is of course necessary to insure good work. Another form of bevel shank cutter is shown at "*C*." The thread is of coarse pitch and fitted loosely to compensate for irregularities due to hardening. The bevel and teeth are finished concentric by grinding. In re-sharpening these cutters, they should be located from the bevel in a holder provided for this purpose. If held between centers, the teeth will not be concentric with the bevel if the cutter happens to be sprung a little, which is generally the case with cutters that have been in use

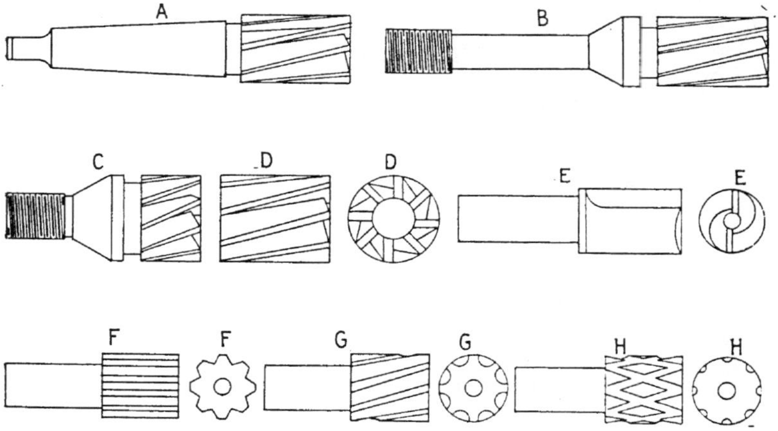


FIG. 53

for some time. A shell cutter is shown at "*D.*" These are easily and cheaply made but they are not reliable as the thread that holds the cutter in position weakens both the cutter and its arbor. A double fly-tooth cutter is shown at "*E.*" These are sometimes used for roughing out cam paths. It is claimed that this design allows the chips to work away from the cut readily. On the other hand, the fact that there are only two teeth is taken by many mechanics as a decided objection.

A teazel, or finishing cutter, is shown at "F." It is a well known fact that ordinary cutters when used for finishing wear undersize in a very short time. For this reason the use of teazels is recommended on accurate cam paths. As the teeth of the teazel are not relieved, they hold their size for a long time. Teazels can be straight or spiral and both forms will give good results. A spiral cut teazel is shown at *"G."* The teazel shown at *"H"* is double cut, this design being popular in some manufacturing centers.

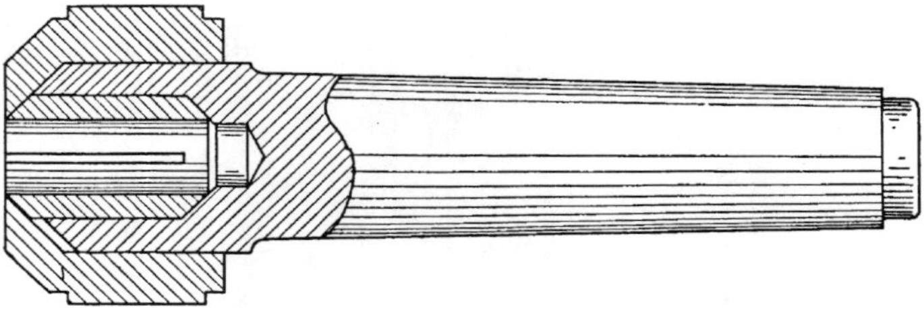

FIG. 54

Any mechanic who has attempted to cut accurate cam paths in cast iron has experienced the annoyance that develops while taking the finishing cut with an ordinary spiral cutter. If the cutter is not perfectly ground it will cause chattering, and, even under ideal conditions, it will sometimes crawl, thus making the path over-size. These difficulties are wholly overcome by the use of teazels for finishing cuts. A teazel removes stock by scraping and, therefore, it should not be used directly after a roughing cutter, but after a finishing cutter that is from 0.002″ to 0.005″ under size.

An efficient holder for straight-shank cutters is shown in Fig. 54 and Fig. 55, while Fig. 56 shows a few cutters and teazels that are used in connection with it. The shell *"A,"* which is split in four places, is compressed by the pressure brought to bear on its ends by the nut *"B."* Thus it grips the cutter shank firmly. Several fillers of different inside diameters are provided for various sized cutter shanks. One of these fillers is shown at *"C."* This cutter holder should be accurately made, and the

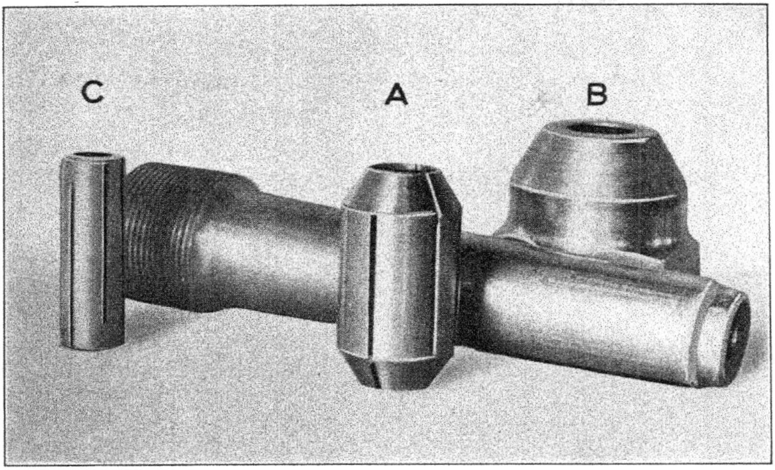


Fig. 55

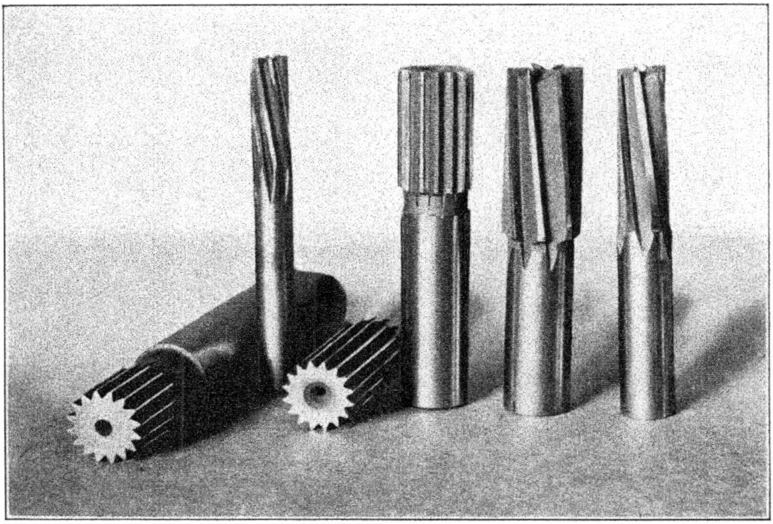

Fig. 56

shell and fillers hardened in oil after which they are finished by lapping and grinding. If used with a reasonable amount of care, a holder of this kind will remain accurate for a long time.

A leader is sometimes spoken of as a permanent master cam and for this reason care should be exercised in cutting it, otherwise a false contour will result. It is, of course, understood that the guide roll and cam cutting spindle are on the same plane, and the first step in cutting a leader is to see that this plane is in line with the center of the milling machine spindle, in cases where a milling machine cam cutting attachment is used. For this purpose the device illustrated in Fig. 57 can be used to good advantage. The button *"A"* should be of the same size as the guide roll. It is fastened to the taper shank by the screw *"C."* The shank fits the milling machine spindle. After placing the shank in position the button is trued up by means of a test indicator. Then the indicator is again used to set the button and guide roll in the same plane. The reading from the button and guide roll should correspond when the milling machine knee is locked in position by tightening the gibs.

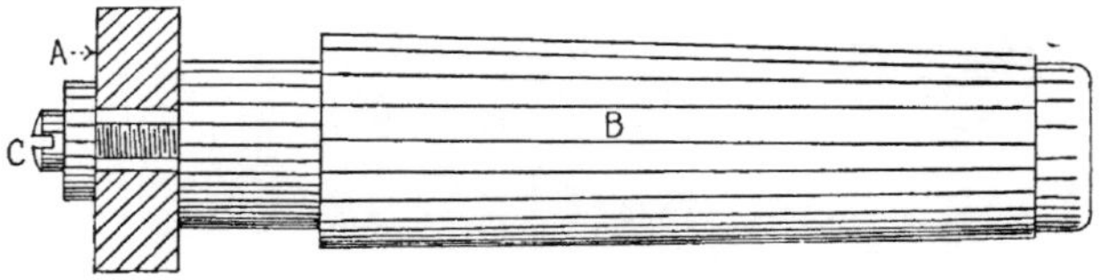


FIG. 57

The locating device is now removed and the roughing cutters, which should be somewhat smaller than the guide roll that is to be used with the finished leader, is placed in position. The hand-made master cam is now placed in position on the cam cutter head spindle with its periphery bearing on the guide roll. The leader blank, which has been previously turned and faced, is placed in the position that is to be occupied by the cams. The end of the roughing cutter is now brought to bear slightly on the face of the leader blank and the cam cutter spindle given one revolution. This marks the desired outline on the leader. The leader is now removed, and the superflous stock removed by drilling and filing. It is again placed in position and the roughing cutter used to bring it nearly to size. The cutter used for finishing should correspond in size with the guide roll and it should run true. A few light cuts with this cutter completes the operation. The leader should be plainly stamped with the size of roll that is to be used in connection with it, also the side that is to go next to the shoulder of the cam cutter spindle should be plainly marked. This is to avoid confusion and to prevent cutting cams backwards. A leader for a drum cam is cut in practically the same way.

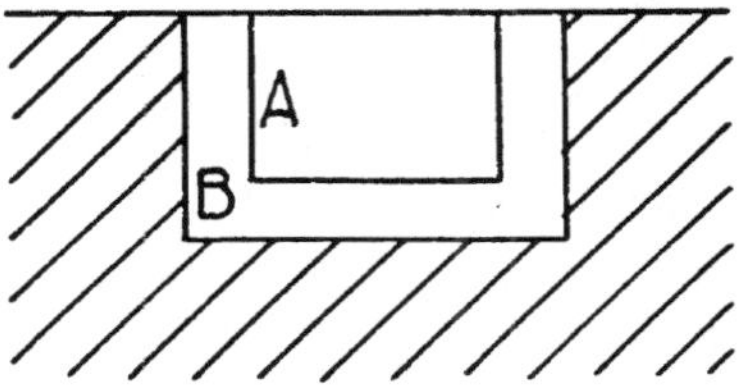

Fig. 58

Face cams are sometimes cut from solid blanks, but this is poor practice. Take the path shown in Fig. 58 as an example. Here it would be necessary to take four cuts to accomplish the desired result. Two cuts would be necessary to remove the metal shown at "*A*" and "B," and afterward another cut followed by a teazel cut would be necessary. By casting the path in the cam at least one of the roughing cuts would be eliminated.

With cams in which the rough paths are cast it is necessary to provide a means of preserving the relation between the leader and the cam blank. A pin as shown at "*B*," Fig. 59, will give good results. This is fastened to the shoulder of the cam-cutter arbor by means of a drive fit. It should be of sufficient diameter to avoid shearing off and slightly undersize to allow it to enter the locating hole in the cam. As both the leader and cam are located by the same pin it is necessary that the locat-

FIG. 59

ing holes correspond, otherwise the contour on the leader and the cast path in the cam will not be in the correct relation. A good means for locating the hole in the cam correctly is to have a spot cast on it for starting the drill; the drilling to be done with the simple jig shown at *"C."* The jig is held in position by means of the plug *"D,"* which fits the center hole of the cam, the spot referred to serving to start the drill in the proper place.

Many cams are made with multiple paths as shown at *"A."* With these cams it is necessary that the locating holes in the leaders be in the proper places to give the correct timing in the finished cam. The locating hole in the cam must also be correctly placed. To accomplish this the following method will prove satisfactory. As the relation between the two paths in the cam *"A"* is known, the two leaders should be set one over the other in relation to give the proper timing. This can be derived from the original cam chart as laid out by the draftsman. The two leaders are now drilled together to fit the locating pin. The next step is to make a metal pattern from which the cam castings are to be made. A plain blank is used for this purpose. After turning and facing to the desired dimensions a locating hole is drilled with the jig previously mentioned. The blank is now placed in position on the cam cutter and the paths cut. The paths should be undersize enough to allow stock for finishing, the amount to leave depending on the dimensions of the cam. In spacing the paths the correct distance from the center, allowance for shrinkage should be made. The paths should be cut with a tapered cutter to allow for draft. This saves profanity on the part of the molder and insures clean castings. The locating hole is now plugged up and a spot made in the same place. This spot is for starting the drill as previously described.

After correct patterns for casting the cams have been made, and leaders for cutting them provided, the operation of cutting cams is quite simple and good results will be obtained if the few important points that follow are observed. Some of these suggestions have been previously referred to but it will do no harm to mention them again.

The machine should be properly oiled; not when the operator happens to feel like it, but at regular intervals. Once a day is not too often when the machine is in constant use. If the operator allows the ways of the sliding-table type of machine to become dry faulty cams are liable to be the result.

Never use several hundred pounds of iron for a weight when fifty pounds are sufficient. Fig. 49 gives a good illustration of this error. Excessive weight throws an undue strain on the attachment as the guide roll is always well to one side. Again, the combination of excessive weight and constant rolling contact is liable to wear the leader out of true contour.

Make sure that the leader is placed in the correct position; for if placed on backwards, spoiled work will result. Leaders should always be marked to show which side goes next to the shoulder; not in meaningless hieroglyphics that are understood by the operator only, but in plain English that can be read at a glance. Care should also be used to see that the guide roll is of the correct diameter, as a variation here will generate an incorrect motion.

Don't drive the cam in position. If it does not fit, there is, in all probability, a reamer in the tool room that will correct the error. Driving cams on and off is a waste of time, and generally ends in springing the arbor out of shape. A sprung arbor will often prove the means of rendering valueless several hundred cams before the error is detected.

Use sharp cutters and make sure that they run true. When cutters run out of true, an undue strain is forced on the few teeth that do the cutting. Finishing cutters should run true, for if they run out the path will be larger at the bottom than it should be. Never attempt to cut a full path from solid stock at one cut as chattering and a broken cutter will result.

In cases where very accurate paths are called for, a teazel should be used for the final finishing, but do not expect the teazel to do the work of a finishing cutter. Owing to its design, a teazel will not remove more than a few thousandths of an inch at best.

When cutting steel cams use plenty of lubricant, either cutting oil or cutting compound. This should be delivered to the work by means of a drip feed—not from an oil can when the operator happens to think of it. Never attempt to teazel steel cams—this is uncalled for and it is not practicable.

In cutting open cams in cases where accuracy of contour is necessary, use a finishing cutter of the same size as the roll that is to be used with the finished cam as a slight variation will often result in noticeable error.

# CHAPTER VII

## Cam Cutting—Continued

### UNUSUAL CAM PROBLEMS

The following unusual cam problems may be of interest to those seeking general information on the subject of cams. At "*A*" Fig. 60 is shown a sharp rise that is

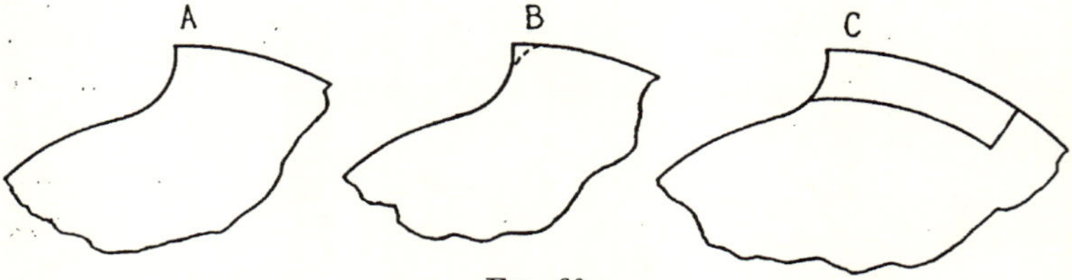


Fig. 60

sometimes used in connection with open cams. The constant action of the roll at this point soon wears the cam as shown by the dotted line in "*B.*" To overcome this objection, the cam is reinforced at this point by means of the hardened piece shown at "*C.*" This piece can be held in place by screws or rivets.

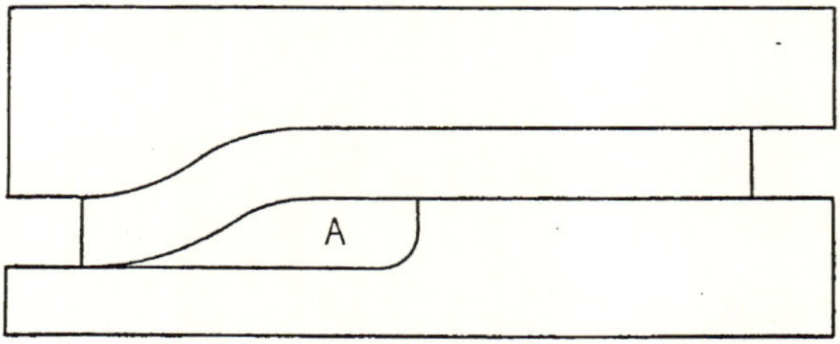


Fig. 61

Path cams are often reinforced on new as well as repair work. An illustration of this is shown in Fig. 61.

As it is not always good practice to make cams of this type of steel, owing to the high production cost thus incurred, it is a good plan to reinforce the portion of the path subjected to the most wear. The piece *"A,"* which is held in position by screws, is made of tool steel, and

FIG. 62

after being fitted in position, it is taken out and hardened in cyanide and oil. These illustrations will, no doubt, bring to the mind of the practical man many instances wherein new as well as worn cams can be reinforced to advantage.

It is sometimes necessary to cut scroll cams as shown in Fig. 62. A cam of this type, which often has a movement of over one revolution, is generally actuated by a rack meshing with a gear on the cam shaft. The master cam is made of brass in the regular way, but it should be reinforced at the back as shown in Fig. 63, otherwise the pressure of the roll would throw it out of shape. The leader is made of cast iron, being cut from the master in the usual way. As it resembles an ordinary path cam,

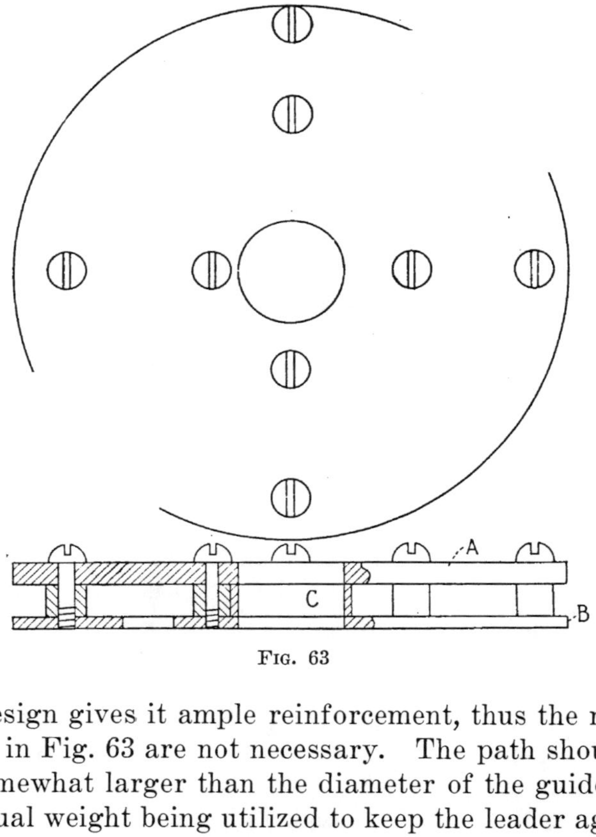


FIG. 63

this design gives it ample reinforcement, thus the means shown in Fig. 63 are not necessary. The path should be cut somewhat larger than the diameter of the guide roll, the usual weight being utilized to keep the leader against the roll which should bear on the inner surface of the path. This principle should also be used with the master cam. The operation of cutting scroll cams has to be watched closely for if the leader were allowed to feed against the guide roll at the end of the path, the spindle or guide roll stud would be sprung out of true.

In Fig. 64 is shown a safety device that can be used to advantage in cutting wiping cams. In cutting cams of this kind it is customary to start the cut at the small

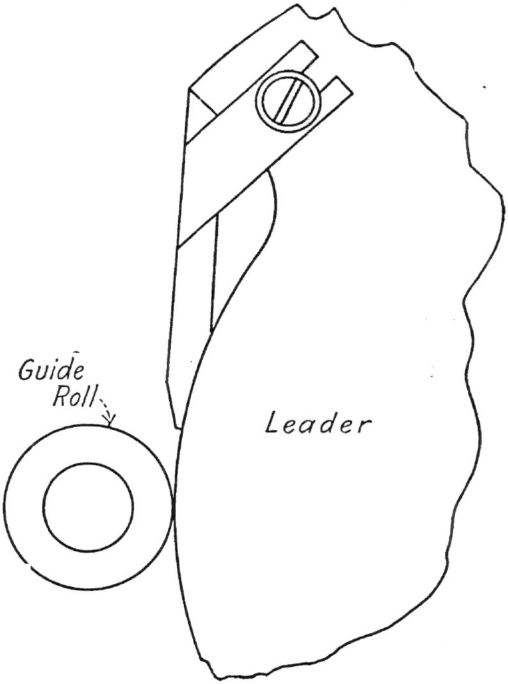


FIG. 64

diameter. The small amount of stock left by the radius of the cutter is removed by another operation. After the leader has run over the roll for a few inches, the piece *"A"* is inserted. This is to avoid accident at the end of the cut as by its means a sudden drop is prevented at the completion of the cut. Without the inserted piece the guide roll and its stud and bracket would receive a strain that might injure them.

A novel cam, sometimes used in sewing machine construction is illustrated in "*A*" Fig. 65. This is a switch cam, so called because the movement is changed every

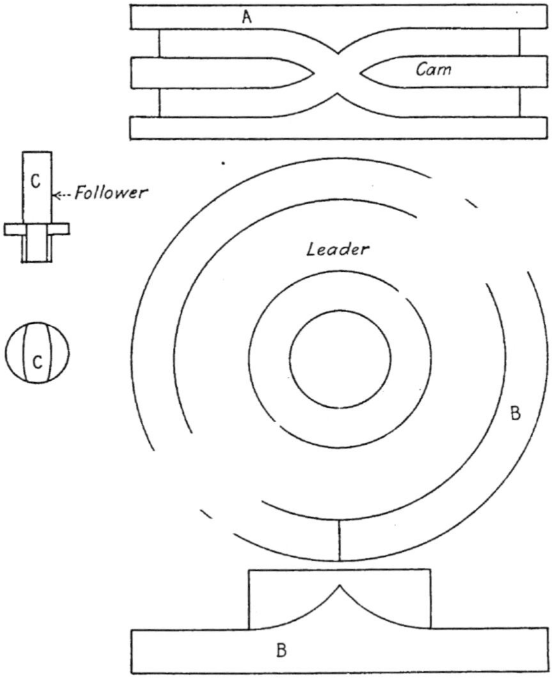


FIG. 65

revolution, two revolutions being necessary to complete the full stroke of the cam. On first thought, it appears that this type of cam would be difficult to cut, but in reality the operation is simple. One leader, as shown at "*B*" is used, and two cuts are made to complete the continuous path. A fixed follower, as shown at "*C*" is used with cams of this kind as a roll follower would be liable to become wedged at the crossing of the path. This movement is inclined to be short-lived, owing to the friction between the path and the fixed follower. Notwithstanding this objection, however, cams of this type can often be used to advantage.

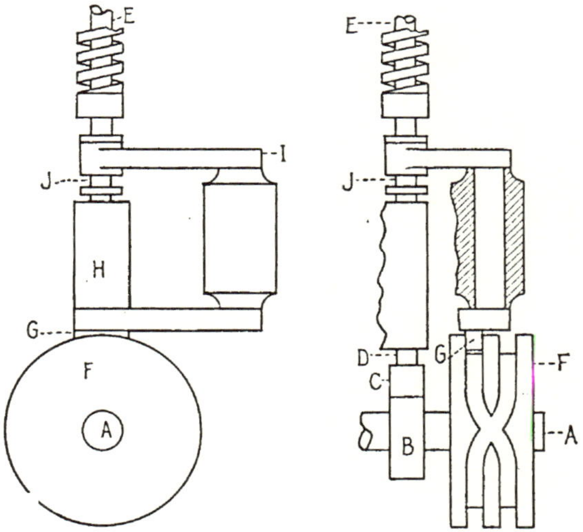


Fig. 66

An application of this cam movement to the valve mechanism of a single-cylinder, four-cycle, internal combustion engine is illustrated in Fig. 66. In this case the switch cam takes the place of the usual half-time shaft. *"A"* is the crank shaft, *"B"* the exhaust cam, *"C"* its follower, *"D"* the push rod, *"E"* the valve stem, *"F"* the switch cam, *"G"* its follower, *"H"* the follower shaft housing, *"I"* the arm that is interposed between the push rod and the valve stem, and *"J"* the sliding button that transmits the motion of the push rod to the valve stem. As the exhaust valve of a four-cycle engine is operated but once every other revolution of the crank shaft, the switch cam provides an end to this means as

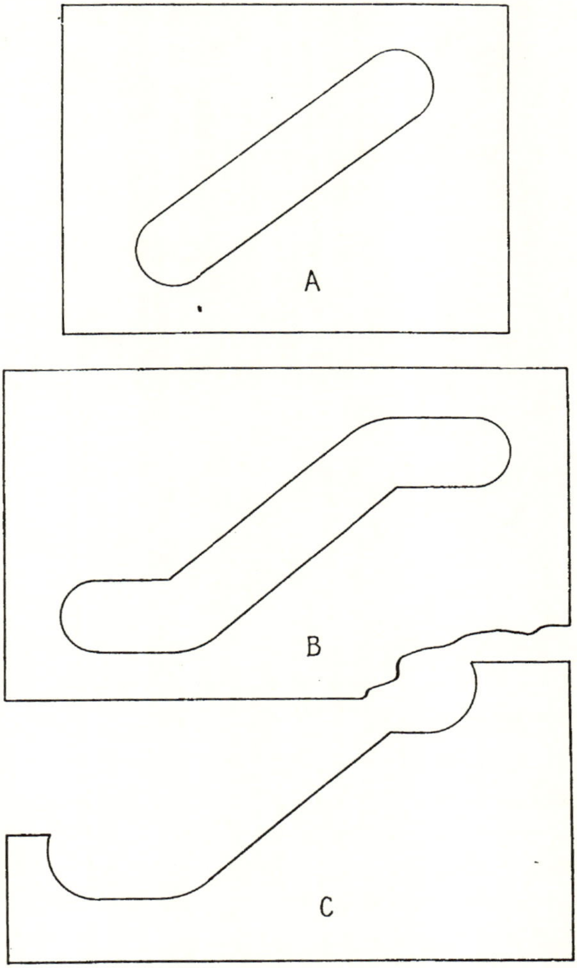


Fig. 67

the button *"J"* is moved from its position between the push rod and valve stem every other revolution, thus preventing the valve from being raised.

A simple plate-cam movement, as shown at *"A"* Fig.

67, can be cut without a leader as the center of the cut is represented by a straight line. With the motion illustrated at *"B,"* in which the movement of the follower is locked before the start and after the completion of the stroke, a leader is necessary to avoid errors of contour, due to cutting back. The leader for this movement is shown at *"G."* In laying this out, the same method is employed as with ordinary cams.

When cutting the cam, this leader is fastened to a disk plate by means of screws and dowels, the disk fitting the spindle of the cam cutter head. The cam to be cut is another disk in the same relative position. The operation of cutting is the same as with any cam, with the addition that care to be exercised to prevent the leader from locking against the roll at the end of the cut.

Is it possible to cut an accurate cam without a leader? This question, which has been discussed for many years in the columns of the technical papers, can be answered in a very few words. Any open cam, of either the drum or disk type can be cut on the universal milling machine with the dividing head, in cases where the motions are to be uniform rises. It is impossible to cut an accurate path cam without a leader, as a properly generated leader is necessary to overcome the factor of cutting back. It is also impossible to cut anything but uniform rises without a leader because the dividing head can not be geared to give anything but a uniform relation between the longitudinal movement of the platen and the dividing head spindle. In rare cases, where uniform rises will answer the purpose, and when a cam cutting attachment is not available, it is considered good practice to utilize the universal milling machine. It must be bourne in mind, however, that this method applies to open cams only.

A movement that can be generated directly on the milling machine is illustrated in Fig. 68. The movement

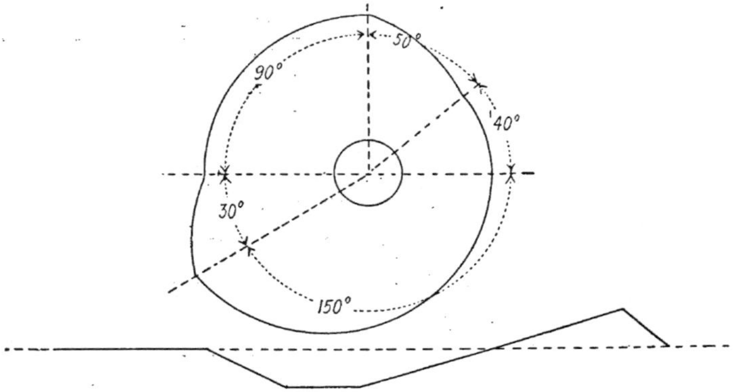


FIG. 68

is as follows: first a rest of 90°, then a drop of ½″ in 50°, followed by a rest of 40°, next a rise of 1″ in 150°, then a drop of ½″ in 30°. After the blank has been turned and faced the movement is laid out and the superfluous stock removed by drilling and filing. The piece is now placed between the dividing head centers, being securely held on a work arbor. The milling should be done with a half round convex cutter of the same radius as the roll that is to be used with the finished cam. The principle of the method in question is shown in Fig. 69.

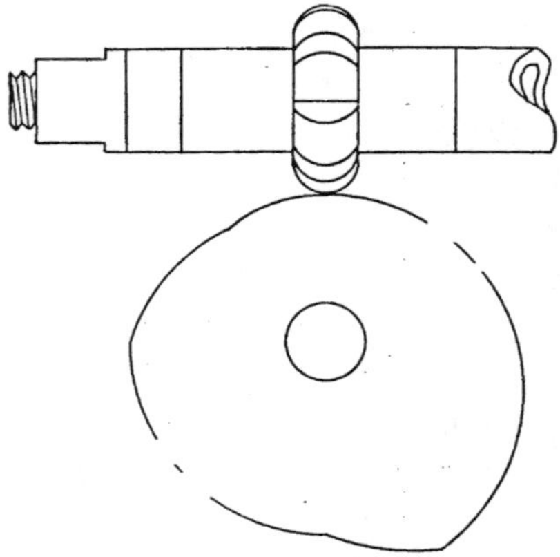

FIG. 69

In setting up the dividing head, a plate having a circle of eighteen holes is used, two holes representing one degree. In milling the cam 360 finishing cuts will be taken. If the cam has been previously filed to say within $\frac{1}{16}''$ of the desired size, roughing cuts can be dispensed with, provided a fine feed is used.

After setting the cutter to the required depth, the first 90° of the contour are milled into 90 cuts. In milling the drop of $\frac{1}{2}''$ in 50° the platen is raised 0.10″ before taking each cut. The rest of 40° is finished by taking 40 cuts without disturbing the set of the platen. In milling the rise of 1″ in 150° the platen is dropped $0.006\frac{2}{3}''$ before each cut. The platen should be dropped below the desired setting and brought to position with an up motion to eliminate errors that might arise through backlash. In finishing the drop of $\frac{1}{2}''$ in 30° the platen is raised $0.16\frac{2}{3}''$ before each cut. This finishes the outline, bringing the cutter to the starting point. The contour of the cam now presents 360 slight ridges which can readily be removed by filling.

Another method of cutting cams without leaders consists of gearing up the dividing head to turn the desired number of degrees while the platen advances the required distance to generate the rise. A drum cam can be held between centers in the usual way. With a disk cam, the dividing head is set at an angle of 90° with the platen, the cam being held on a special arbor. In this case the cutter is driven by a vertical milling attachment. The method in question will give good results if the operator uses extreme care, both in figuring out the leads and in milling the rises. Where there is a cam cutting attachment available it is best to use it, even if only a few cams are wanted. A good mechanic, who thoroughly understands laying out and cutting cams, can often make a leader in the time that would be consumed in figuring out the leads and setting up the milling machine and dividing head for cam cutting by the method just described.

In cases where only a very few cams of the disk type are wanted, they can sometimes be finished to advantage by filing. The large eccentric cams used on river steamboats are never cut on a cam cutter. They are cast from an accurate pattern and finished by just enough filing to insure their peripheries being square. As there is generally $\frac{1}{8}''$ to $\frac{3}{8}''$ allowed for backlash between these cams and their confining plates, or followers, accurate finish is not necessary.

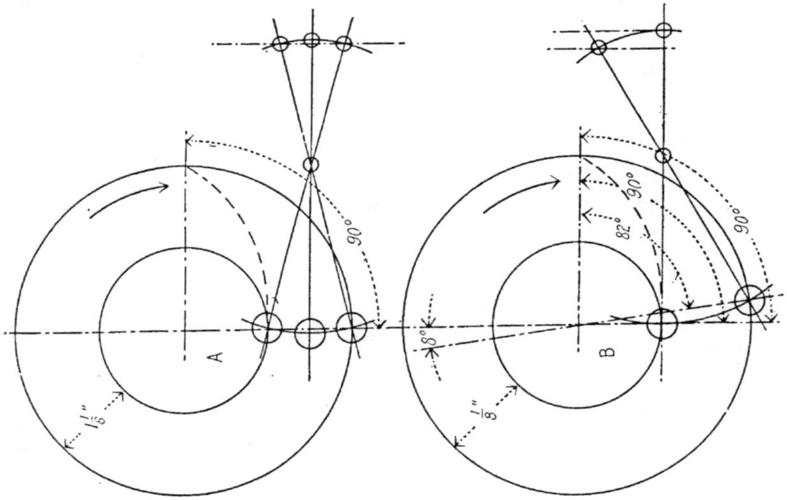


Fig. 70

The cams shown in Fig. 70 have a rise of $1\frac{1}{8}''$ in 90°. Let it be assumed that we wish to alter the time of rising without changing the contour of the cam. It is understood that the best practice is to locate the cam lever fulcrum stud midway in the stroke, as shown in illustration *"A."* While the roll and the end of the cam lever both move in arcs, as the illustration shows, the actual timing, as cut on the cam contour, is accurately conveyed to the machine parts that are actuated. If, for experimental or other purposes, it is desirable to alter the timing slightly without changing the cam, we can accom-

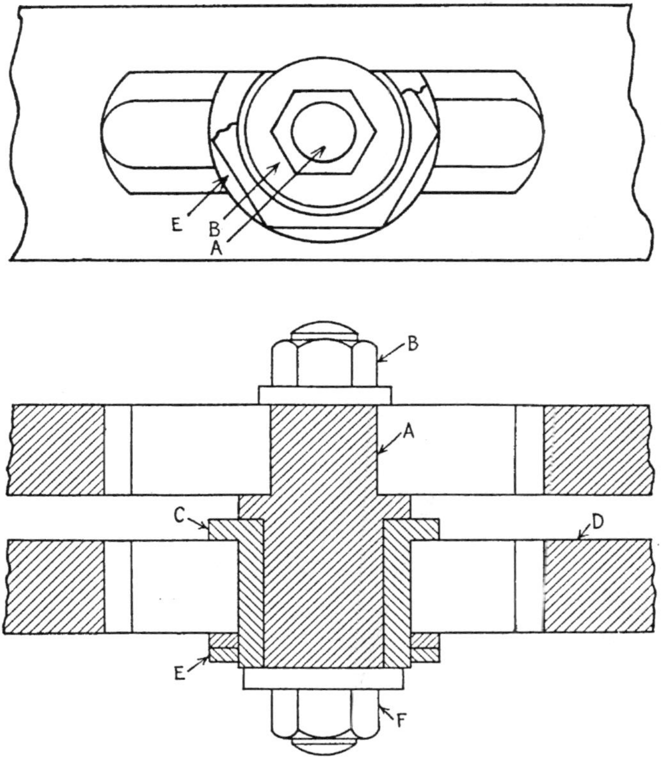


FIG. 71

plish this by offsetting the fulcrum stud as shown at *"B."* Here the fulcrum is placed on a line with the start of the cam rise. As the lever moves over its fulcrum it draws the roll away from the center line of the cam, thus the cam stroke in this particular case is completed in 82° instead of 90°, as originally laid out. Thus the motion is shortened by 8°. This practice incurs a factor that is sometimes detrimental as the end of the lever moves to one side of the line of converted motion. If the cam were rotated in the opposite direction of that shown by the arrow, the movement would be prolonged; that is, the stroke would be completed in 98°. Offsetting the fulcrum stud is sometimes advantageous while experimenting with cam movements, although it cannot be called good practice in the strictest sense of the word.

The machine designer is occasionally confronted with the problem of providing a cam movement with a variable rise. This can be accomplished by using the adjustable fulcrum stud illustrated in Fig. 71. The stud "*A*," which is slightly flattened on two sides, is held in position in its slot by the nut "*B*." The stud bushing "*C*" is also flattened, being locked in position in a slot in the cam lever "*D*" by means of the nut "*E*." The bushing is held in position on the stud by the nut "F." When this nut is tightened the bushing must, of course, have slight clearance to move freely on the stud. In changing the stroke, the two slots are first brought in line, as shown in the illustration, and the nuts "*B*" and "*E*" loosened. Then the stud is moved to a position in the slots to impart the desired stroke, after which the nuts are tightened. This feature is simple in construction and efficient in results.

Sometimes occasions arise wherein it is desirable to provide a cam having a variable timing. To accomplish this an adjustable cam is necessary A cam of this type is illustrated in Fig. 72. The movement consists of a rest period of 180° followed by a drop of 1″ in $52\frac{1}{2}$°, a rest of 75°, and a rise of 1″ in $52\frac{1}{2}$°. In this case the problem is to provide means to shorten the 75° period of rest. The cam is made in two pieces, the cam proper being provided with a hub over which the supplementary piece rotates. The two sections are clamped together by a stud and nut. The stud is fastened to the supplementary piece by means of a drive fit, moving in the slot in the cam proper. Both sections are cut at one operation, the stud being clamped in the position in the illus-

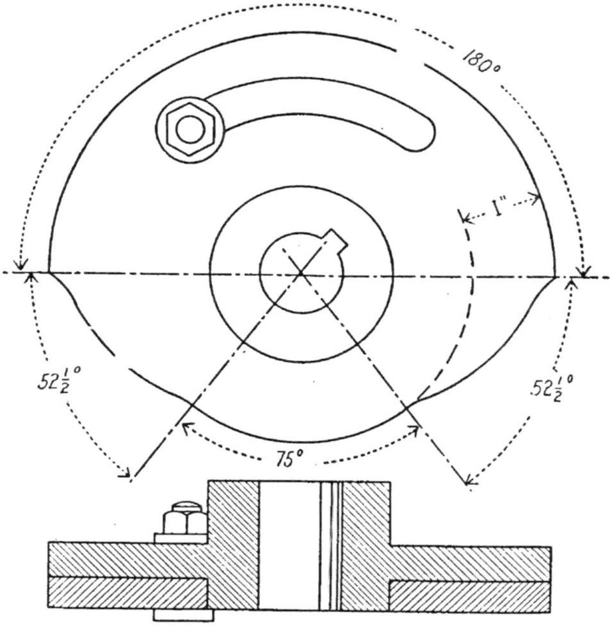


FIG. 72

tration. To shorten the stroke, the stud nut is loosened and the cam rises brought toward each other to the desired position where they are held by tightening the stud nut. One roll, long enough to cover both sections, is used with this cam. The cam is shown at its full period of rest of 75°. By closing up the cam, this rest period can be eliminated entirely. This, of course would lengthen the long period of rest by 75°. It is possible to construct a great variety of adjustable cams but the one described will serve to illustrate the principle.

Two types of cam rockers are shown in Fig. 73. As before stated, one use of these is to lift the valves of the beam engines used on side-wheel steamboats. The movement is driven by an eccentric on the paddle-wheel shaft

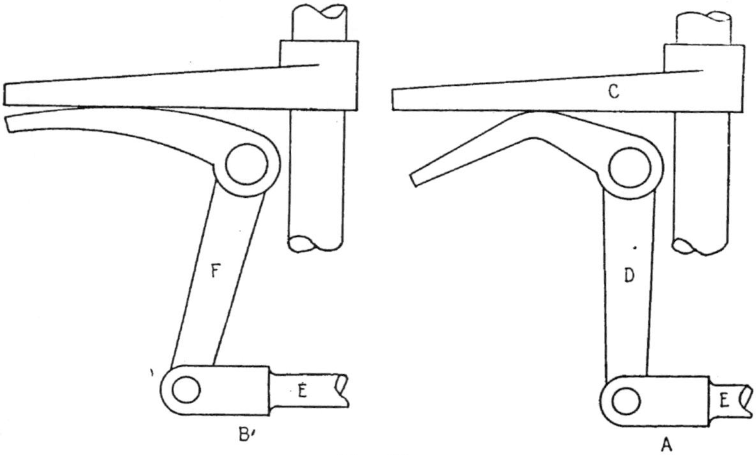


FIG. 73

through the medium of the rod *"E."* There are many methods in use for generating the curves on these cam rockers, all of which, no doubt, possess advantages. The following methods, which are very simple have often been used to advantage. The movement shown at *"A"* gives the toe piece *"C"* a rist of 2″ while the lever *"D"* travels through an arc of 45°. To lay out this movement, a chart, as shown in *"A"* Fig. 74, is drawn. This represents the rise of 2″ divided into eight equal parts by the lines 1, 2, 3, 4, 5, 6, 7, and 8. Dimension *"B"* corresponds to the distance from the rock shaft to the rod carrying the toe piece, dimension *"C"* corresponds to the radius of the rocker hub, while dimension *"D"* equals the distance from the center of the toe-piece rod to the middle of the toe piece. *"E"* is a template made of heavy paper. Dimensions *"H"* and *"C"* correspond. The desired angle of 45° is included between lines "0" and "8" on the template. This angle is divided into eight equal parts by the lines 0, 1, 2, 3, 4, 5, 6, 7, and 8. The template is fastened in its correct relative position by means of a thumb tack, the top line "F" coinciding with line "0" on the chart. The template is now moved until its line "1" is opposite line *"G."* Line "1" on

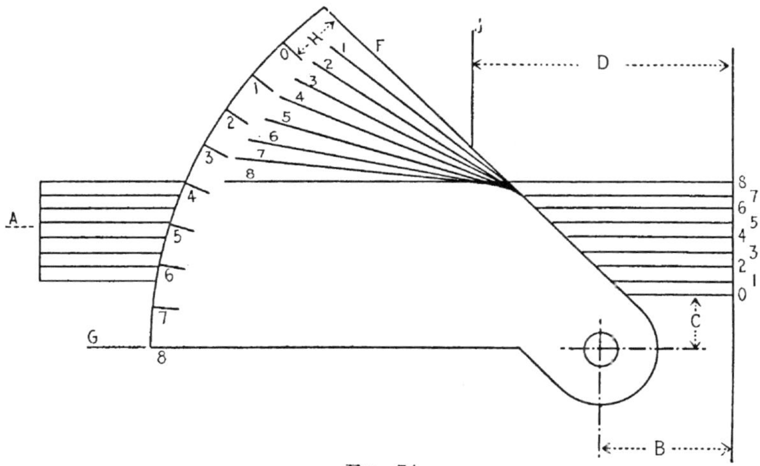


FIG. 74

the chart is now continued over the template. The template is again moved until lines "2" and *"G"* coincide and line "2" of the chart carried over the template. This process is continued until line "8" is drawn. The template is now in the position shown in the illustration. It is seen that a curve has been generated on the template. This is the desired curve for the rocker.

While the above is an easy working motion it possesses a disadvantage as practically all the bearing on the toe piece is in the vicinity of the line *"J."* This can be overcome by using the rocker shown at *"B,"* Fig. 73. This rocker has a curve that bears on the toe piece with a long, rocking motion. To lay out this curve we can proceed as shown in Fig. 75. The lines *"A A"* equal the desired rise, dimensions *"B* and *C"* being in the

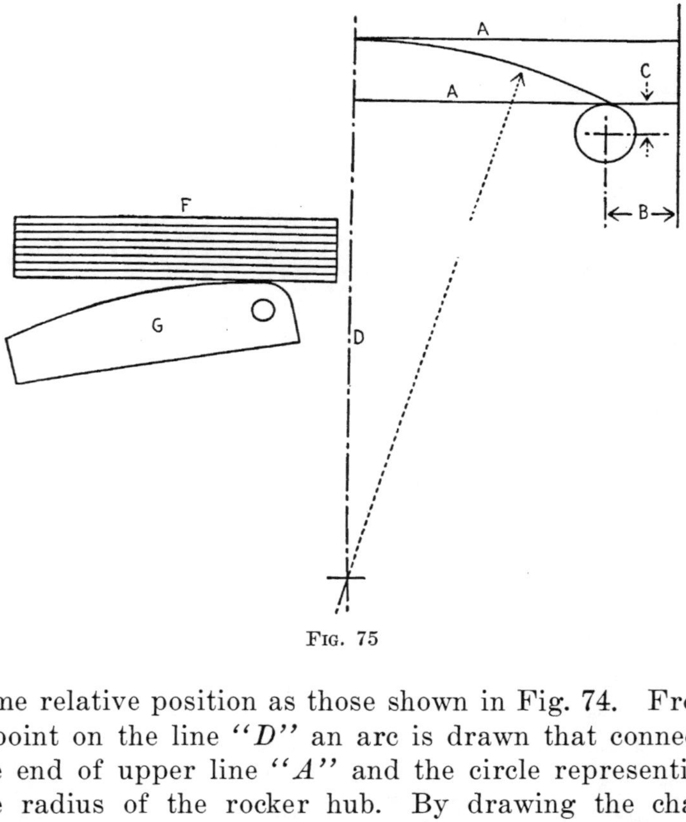


FIG. 75

same relative position as those shown in Fig. 74. From a point on the line *"D"* an arc is drawn that connects the end of upper line *"A"* and the circle representing the radius of the rocker hub. By drawing the chart *"F,"* and making the template *"G"* which is pivoted in the correct position by means of a thumb tack, it is a simple matter to demonstrate that the bearing point moves along the toe piece as it is raised. In this case, the point of applied power is changed as the toe piece rises, and as the leverage is decreased in like proportion, more power is exerted at the completion of the stroke than at the beginning. To avoid throwing an undue strain on the eccentric and its rod, the lever "F," Fig. 73, should at least equal the distance from the end of the toe piece to the point directly over the rocker shaft.

Cam rockers of the kind in question are never cut on a cam cutter. The type shown at "*A*," Fig. 73, is laid out by means of a sheet metal template, made according to the method described, and generally finished by filing, care being exercised to make sure that the bearing surface is square. As the rocker shown at "*B*," Fig. 73, is nothing more or less than an arc, the piece can be strapped to the face plate of the lathe, or the platen of the boring mill, and the correct radius finished by turning.

# CHAPTER VIII

## Cam Grinding

Some years ago, in the days of fine-grit, close-bonded emery wheels, cam grinding was practically unknown. At the present time, however, owing to the development of grinding machines and grinding wheels, cam grinding has become common practice, especially in the automobile industry. Any open cam, whether hard or soft, can be finished accurately by grinding if its surface does not present concavaties or less radius than that of the grinding wheel used.

The attachments, or fixtures, for cam grinding can be divided into two distinct types: Those employing regular disk wheels, and those calling for the use of cup or cylinder wheels. Where extreme accuracy as regards contour is desired, it is necessary to use a cup or cylinder wheel, for when a disk wheel is used the contour of the cams that are ground is bound to change slightly as the wheel wears away. This change is not very noticeable, and it is of little importance with automobile engine cams. With eccentric cams, however, where it is necessary for the cam to fill the space between the followers at all times, a cup or cylinder wheel is necessary.

The only important argument in favor of the disk wheel for cam grinding is that it cuts faster than a cylinder wheel. This argument is open to exceptions, however, as excellent results as regards cost production can be obtained with the fixture shown in Fig. 76. This fixture, which was designed for use on a vertical spindle grinding machine, consists principally of a base on which is mounted a swing-frame for carrying the leader and

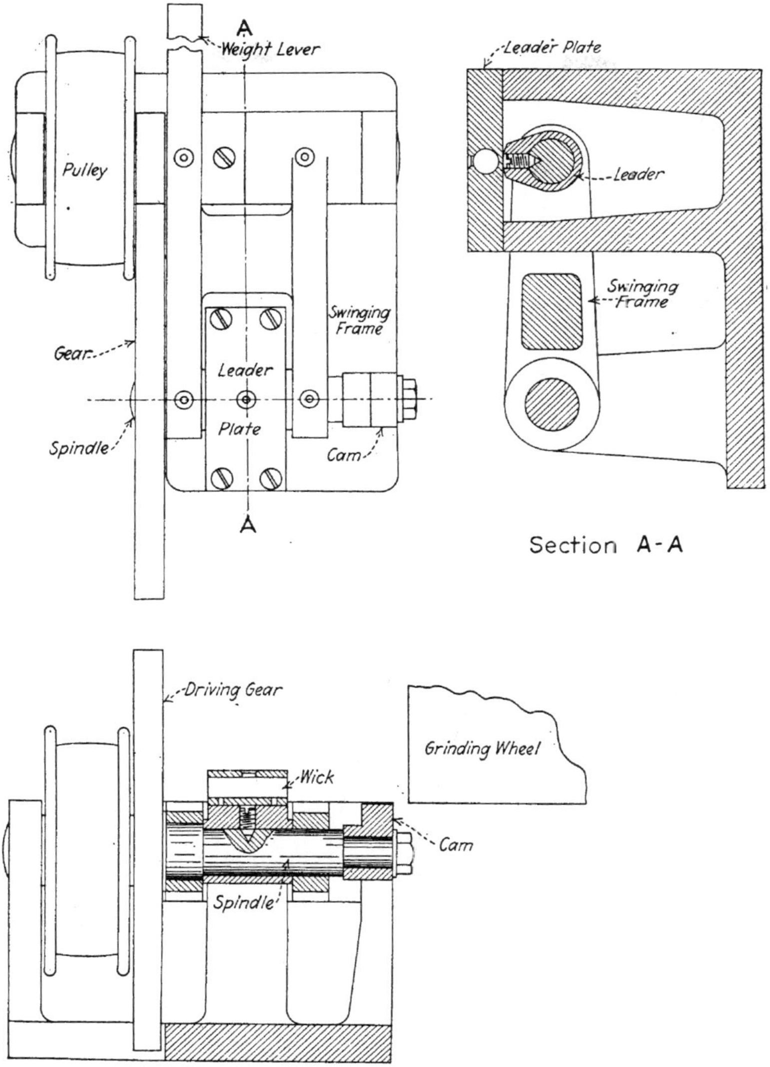
A
Weight Lever
Pulley
Swinging Frame
Gear
Leader
Plate
Spindle
Cam
A
Leader Plate
Leader
Swinging Frame
Section A-A
Driving Gear
Grinding Wheel
Wick
Cam
Spindle

FIG. 76

cam spindle. This spindle is driven by a spur gear which in turn is driven by a pulley through the medium of a pinion. A hardened and ground plate takes the place of the usual guide roll, the leader being held in contact with this plate by means of a lever and weight. The plate is provided with an oiling device consisting of a chamber packed with wick, or felt. This keeps the face of the leader lubricated at all times, thus eliminating unnecessary friction.

While in use, this attachment should be protected by a sheet metal cover, otherwise the centrifugal action of the grinding wheel would throw grit into the working parts. This fixture is simple and practical, and, if properly constructed, it will give entire satisfaction. It possesses one slight disadvantage—that is, the surface speed of the spindle is alternately retarded and accelerated by the planetary action of the driving gear over its pinion. This, however, is not a serious detriment.

The grinding wheel should be of a coarse combination grit and soft grade. For roughing out cams from soft stock a combination grit with 16 for a base, or a straight 24 grit will give good results. For the finishing of cams after they have been hardened, the wheel should be of a softer grade and somewhat finer in grit—20 or 24 for the base of the combination. A combination grit will give the best results for the finishing operation. These grits may seem coarse to one who is not conversant with the fundamental principles of grinding with a high speed cylinder wheel. In this case it is the combination of coarse grit, soft grade, and high wheel speed that gives the desired results.

While grinding, the cam should be rotated very slowly —8 to 15 revolutions per minute. A medium traverse feed should be used—$\frac{1}{16}''$ advance for each revolution of the work. While roughing out cams from soft stock, the wheel should be fed down as much as the work will stand at each complete stroke of the platen. This has to be determined by experiment as local conditions have to be given consideration in practically every case. For the finishing operation, after the cams have been hardened, a few light cuts only are necessary. Under ordinary conditions, 0.010″ should be an ample allowance. When grinding cams that are to be held in position of their shafts by means of a tapered pin, a locating key provided with a teat should be used. The key fits a slot milled in the spindle for this purpose while the teat enters the hole to be occupied by the tapered pin. If a tapered pin were used for locating the cams on the fixture, the spindle would become sprung in time, through driving the pin in position.

The original master cam is left soft and is used for finishing a permanent hardened leader. The fixture in question can be depended upon to finish with absolute accuracy as regards contour. When the leader becomes worn, a new one should be substituted. For finishing new leaders from time to time, a hardened master, ground from the original soft master should be preserved. This master should be plainly marked as a means of identification. It is also necessary occasionally to face off the hardened plate as it becomes worn in time through contact with the leader.

A fixture for grinding accurate eccentric cams is illustrated in Fig. 77. The cams in question enter into the construction of shoe machinery. The cam and its leader are both mounted on an arbor that is held between centers, this arbor being driven by means of a worm gear and worm. In this particular instance the cam is held in position by means of a tapered pin. The fixture shown is of the swing type, with the guide roll on the side away from the grinding wheel. In this case a concaved leader is necessary to generate a convex cam. When the leader becomes worn it should, of course, be re-finished. The method employed for this operation will be explained later. It is claimed that the reversed, or concaved leader is a very efficient design as it tends to eliminate chatter-

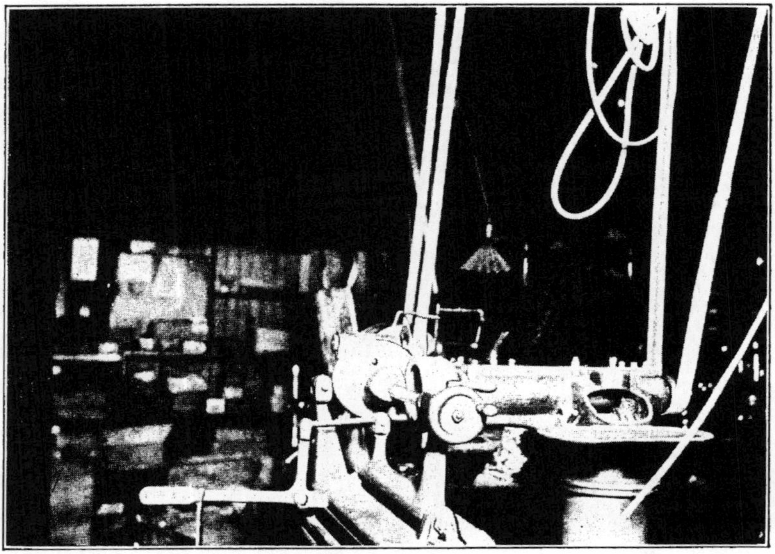

Fig. 77

ing without bringing an undue strain on the guide roll and leader, which is sometimes the case when a heavy weight is used.

The cams in question, two of which are shown in Fig. 78, are made of steel castings, case hardened. In grinding them, a slow work speed is used—10 revolutions per minute. Approximately 0.01″ is left for finishing. The platen is actuated by an automatic feed, advancing $\frac{1}{64}$″ for each revolution of the cam. The wheel is 46 grit, *K* grade, and is run at a surface speed of 5,000 feet per minute. As the illustration shows, the work is ground dry which curtails production to some extent.

While the process of grinding these cams is simple and readily comprehended by any mechanic, the method used in generating the leader is novel and worthy of consideration. By referring to Fig. 79 it is seen that the grinding wheel is removed and that a guiding device is fastened to the end of the wheel spindle. The outer plate of this device is adjustable and is set to bear on the face of the master cam with medium pressure. This master

FIG. 78

cam, which is accurately made, fills the guide at all points as it revolves. With the master cam and leader blank in position on the arbor the desired outline is scribed on the blank by means of a surface gauge. One revolution of the leader accomplishes this. The scriber of the surface gauge should be set central with the arbor. While scribing this line the roll bracket is removed.

The leader is now removed, filled approximately to the desired shape, and again placed in position. A teazel made of high-speed steel is now placed in the position occupied by the guide roll. This teazel is driven by an overhead belt as the illustration shows. As the swing movement of the fixture is now controlled by the master cam revolving between two fixed plates, the leader can

FIG. 79

be brought in contact with the teazel by means of the cross feed. By taking a few light cuts, the outline of the leader is made to correspond to the motion generated by the master. It is necessary for the teazel to be of the same diameter as the roll used in connection with the leader. In event of the leader becoming worn out of shape through use, it can be corrected readily by the method used in finishing a new leader. In Fig. 80 is shown the teazel, which is made integral with its arbor, the guiding device, and the arbor on which is mounted the master cam and leader.

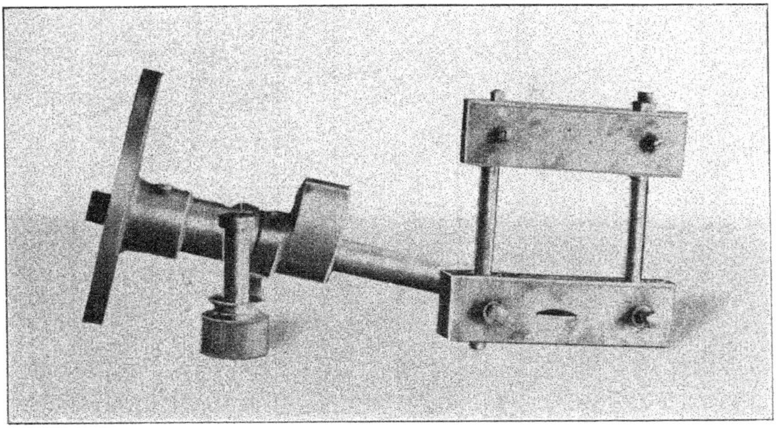

FIG. 80

A simple device for grinding automobile engine cams on a plain or universal grinding machine is illustrated in Fig. 81. The principle is practically the same as with the fixture just described; a weight holding a reversed motion leader against the roll. When this principle is employed, the pressure of grinding only, which is determined by the cross feed, is brought to bear on the cam. This device is simple to construct and will give good results. The spindle is driven by a gear and pinion, power being conveyed by means of an overhead belt. Two leaders, combined as one unit, are fastened directly to the spindle. One leader is for inlet cams and the other for exhaust cams. The guide roll bracket is provided with a sliding base, thus the change from one leader to the other can be instantly made. This attachment is designed on the swing principle and it is so simple that the details are readily understood from the illustrations.

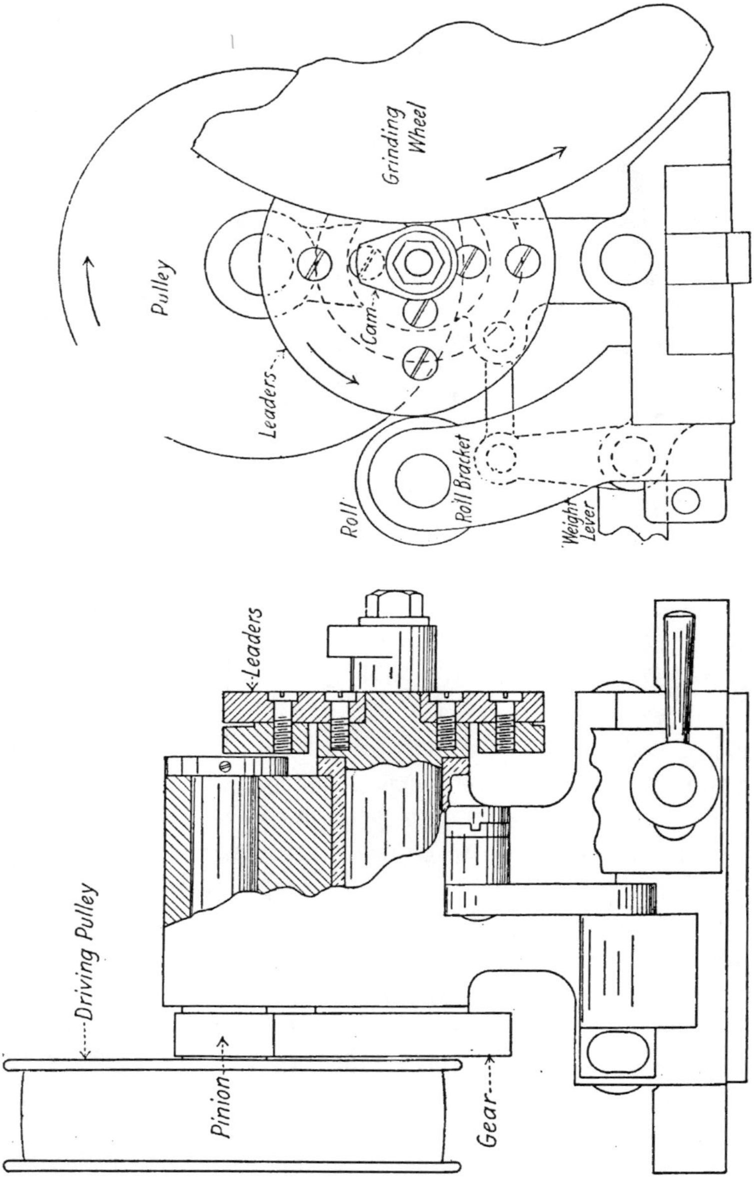
Grinding Wheel
Pulley
Cam
Leaders
Roll
Roll Bracket
Weight Lever
Leaders
Driving Pulley
Pinion
Gear

FIG. 81

With any cam grinding fixture it is necessary to provide means for shaping the leaders accurately. With the fixture in question, the leaders are generated on the milling machine; the principle employed being shown in Fig. 82. The master cam is held in contact with the master plate by means of a weight applied to the lever.

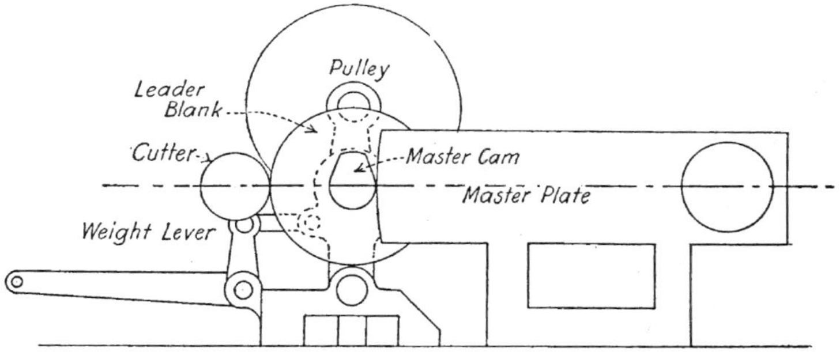


FIG. 82

As the motion is the reverse of that used while grinding, a cord running over an overhead pulley has to be used. The master cam generates the motion over the master plate, while the milling cutter, which is of the same size as the guide roll, mills the correct contour on the leader. While milling the leaders, the attachment should be fed slowly by means of an overhead belt connected to the driving pulley. The master plate is made of cast iron, the contour being turned to the diameter of the grinding wheel. The hole in the base is for convenience in strapping to the milling maching platen.

Where extreme accuracy is desired, the master plate can be re-turned to different radii to correspond to different wheel diameters. Thus, if leaders were to be generated to be used in connection with wheels of 18, 17, 16, 15, and 14 inches, we would have five sets of leaders. After generating the first set for the 18″ wheel, the master plate would be turned to a radius of $8\frac{1}{2}''$, to correspond with the 17″ wheel, and another set of leaders milled. Next would come the leaders for the 16″ wheel and so on down to the last set for the 14″ wheel. It is necessary to mark the leaders to designate what size of

FIG. 83

wheel they were generated to match, thus: "For 18″ wheel," etc.

In making the leader for removing the superfluous stock, the end of the cutter can be brought to bear on the side of the leader blank, which is now given one revolution which causes the cutter to mark the desired outline. The superfluous stock can now be removed by drilling and filing, or directly on the milling machine by means of several cuts. In finishing, light cuts should be taken and the leader should be fed slowly. The cutter must be of the same diameter as the roll to be used on the leaders. Also, it must run true and be set to bring it central with the spindle of the attachment, when it is at the highest point of the ark caused by the swing motion.

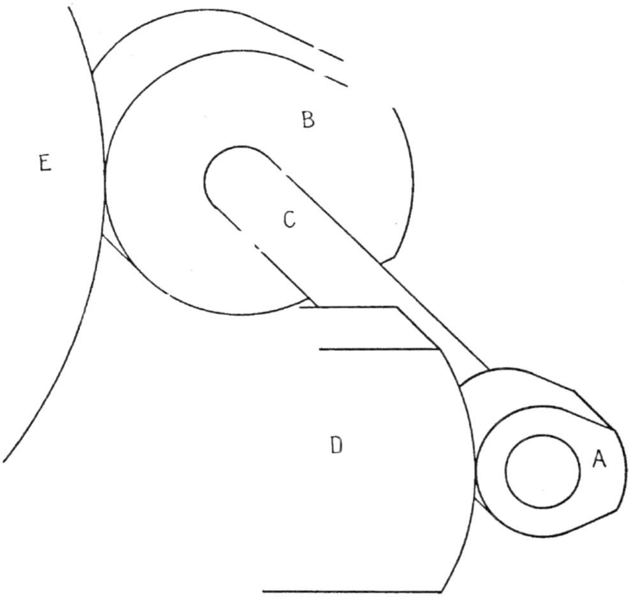


FIG. 84

Another device for grinding individual cams is shown in Fig. 83. This is a product of The Landis Tool Company and is designed for use with their plain and universal grinders. The device is constructed on the swing principle, with the work and the leader on the same spindle. A novel feature of this device is that it grinds its own leaders from actual sized models, or masters. This feature is worth careful consideration for two reasons: It eliminates the roundabout process previously described, and makes possible the construction of accurate hardened leaders. In the cam grinding fixture under consideration the leader operates against a stationary guide, or shoe plate, instead of the customary roll. Several guides are provided having radii to correspond to different wheel diameters. When a new wheel is mounted, say 18″ diameter, a plate to correspond to this diameter

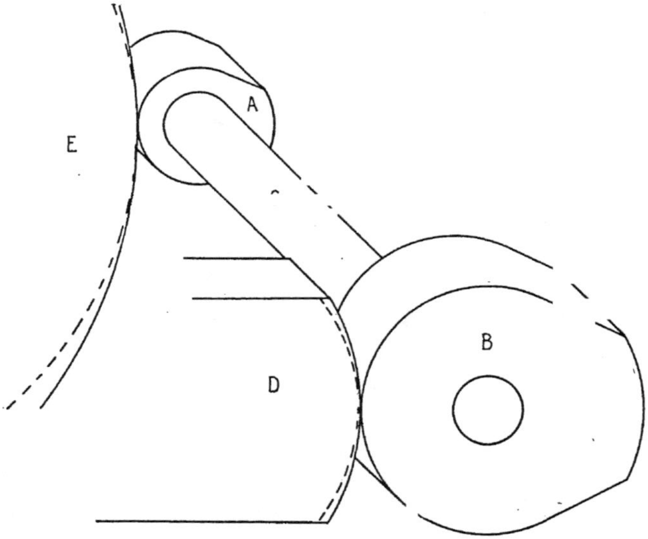


FIG. 85

is used. When the wheel is worn to a diameter of 17″, the 18″ plate is removed and one that corresponds to the 17″ wheel substituted. By this arrangement, errors in contour due to wheel wear are eliminated.

While the general design of the device in question are clearly shown in the illustration, the principle can perhaps be better understood by referring to Fig. 84.

The master cam, "*A*," and the leader to be generated, "*B*," are mounted on the shaft "*C*." The radius of the master plate, "*D*," and the grinding wheel, "*E*," are the same. It is evident that the master, running over the master plate, is bound to impart the desired contour to the leader.

For the actual grinding of the cams, the leader thus generated is placed in the proper position against the master plate as shown in Fig. 85. In this case large leaders are generated from small masters. In commercial cam grinding, however, better results are obtained by employing large leaders to generate small cams. In illustrating the principle used in compensating for wheel wear, let it be assumed that the leader, "B," was generated on an 18″ wheel, the motion being derived from a master running over a master plate of 9″ radius. After the wheel has worn to a diameter of 17″, for instance, as shown by the dotted line, it is evident that a very slight change in contour will result. Now, if we substitute a plate having a contour to correspond to the wheel, as the dotted line on the plate shows, it is obvious that we have equalized conditions. It is readily seen that the principle is not unlike that illustrated in the device for use with the cylinder wheel as shown in Fig. 76. In the first instance we work on planes, while in the second case, radii are substituted.

With the development of multiple-cylinder automobile and airplane engines a demand arose for a grinding device for finishing cams made integral with the cam shaft. A means to this end is illustrated in the attachment shown in Fig. 86. This is also a product of The Landis Tool Company, being designed for use on their plain and universal grinders.

This unit was designed to produce the necessary cam contours demanded by automobile engineers for function-

Fig. 86

ing the valves of internal combustion engines with a high degree of efficiency. The swinging bracket that contains the master cams and their spindle is of tubular cross section being supported directly under the master cams and work centers, close to the machine bed. The object of this construction is to balance the swing bracket so that strains are eliminated. As the work is carried directly over the fulcrum center, vertical movement and the change of grinding contact on the wheel periphery is reduced to a minimum. Thus the difference in cam contour caused by wheel wear is slight. The master cam is mounted directly on the headstock spindle in line with the work so that lost motion is eliminated. The master cam is kept in contact with its roll through the medium of compression springs, the tension being adjustable to suit varying conditions.

Accurate work is assured when the grinding wheel is of a diameter equal to that of the roll used in contact with the model cam used to generate the contour on the master cam. Errors in cam grinding due to wheel wear are caused by the contact between the work and the grinding wheel traveling above and below the wheel center. The greater the variation of grinding wheel contact from the center, the more pronounced the variation in the contour of cams ground with different sized wheels will be. In the machine illustrated, the work is swung equal distances from a point directly over the fulcrum center so that the work moves practically in a horizontal plane.

In Fig. 87, the camshaft, *A,* is driven by the dog, *B,* which in turn is driven by the driver, *C,* mounted on the spindle, *E.* This driver is adjustable to facilitate setting the camshaft in the correct relation with the master cam. The spindle, *E,* is driven by the worm wheel, *F.* The master cams, *G,* are mounted directly on the spindle by means of a taper and a nut. Roller, *H,* makes contact with the master cams and is moved from cam to cam by the lever, *I*. The relative position of the master cam roll is designated by the pointer, *J*. Through the medium of a cam on the shaft, *K*, the roll is disengaged from the master cam by a half turn of the lever, *I*. The roll is shifted to the next cam by completing the turn of lever *I*.

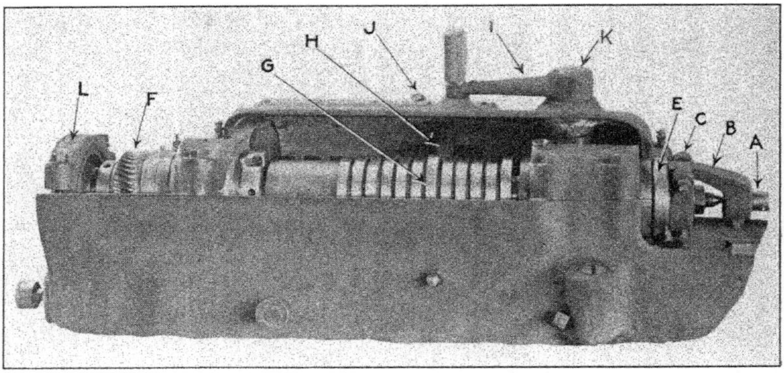


FIG. 87

A brake, L, is provided on the master cam spindle with the object of compensating for any lost motion which would be transferred to the cam as the pressure is reversed when passing over the highest point of contour. A cross section and end elevation of the device is shown in Fig. 88.

The selection of wheels for cam grinding on a cylindrical grinding machine is a simple problem as we have only two factors to consider: Whether we desire to rough out soft cams, removing as much stock as possible in a given length of time, or whether we desire to finish hardened cams, removing only a few thousandths of an inch.

We can take as an example the cams made from bar stock as shown in Fig. 44. As these cams come from the

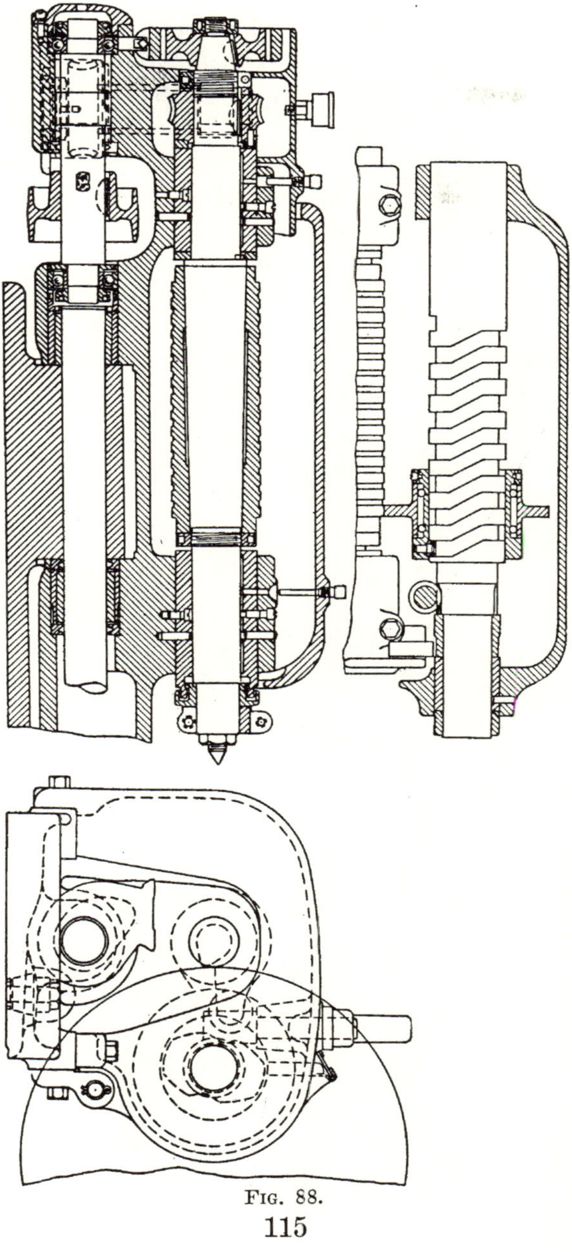

FIG. 88.

turret lathe they are in the form of eccentrics. The desired contour can be formed on the grinder, before the cam is hardened, in a rapid and efficient manner by using a coarse, free-cutting wheel. A 24 straight grit wheel in P grade will give excellent results. This grade may seem rather hard to use on a cylindrical grinder, but it must be taken into consideration that the operation is a roughing one, wherein the wheel must be hard enough to stand up well under a heavy cut. Plenty of water should be used, and the cam fed directly to the wheel. The traverse feed should be discarded altogether. Approximately 0.10″ should be left for the final finishing after hardening.

In grinding individual cams that have been hardened, 24 combination grit, L grade will be found satisfactory. Here, a slight traverse feed should be used in connection with a wide faced wheel. Thus, if we were finishing cams having a 1″ face, a wheel with a 2″ face could be used. The traverse feed should be just enough to allow the edge of the wheel to overlap the edge of the cam a slight distance, say $\frac{1}{16}$″.

Cams made integral with their shafts are drop forged with a liberal allowance for finishing. Sometimes the ends and the sections between the several cams are roughed out in the lathe but equally good results are obtained by using the roughing wheel above referred to, grinding from the rough stock. The same wheel is then used for roughing out the cam contours. These cam shafts are carbonized, or pack hardened, after which they are straightened to correct errors caused by the sudden contraction in the cooling bath, while the piece was red hot. In the final finishing, the ends and the sections between the cams should be ground enough to remove the scale, leaving the work several thousandths of an inch over size. Next the cams are ground to size, and last of all the straight portions are finished. The 24 combination grit, L grade wheel can be used for this purpose.

Occasionally, we find manufacturers who prefer an extra final finishing of the cam contours by hand. In

FIG. 89.

this case a fine-grit abrasive cloth is used. This operation is illustrated in Fig. 89. While a semi-polished surface is imparted it is obvious that the amount of stock removed is practically *nil.* Whether this final finishing is good practice is purely a question of conjecture and a matter of personal taste. Grinding machine and grinding wheel manufacturers inform us that the grinding wheel alone imparts the necessary finish, and, to clinch their argument, they cite numerous automobile manufacturers who follow this practice.

On the other hand, the photograph from which Fig. 89 was reproduced was taken by the author in one of the large automobile manufacturing (not assembling) plants in Detroit, Mich. This final hand finishing, which calls for the services of a skilled workman, imparts a superfine surface, to be sure, but whether this finish is necessary from a practical point of view must be decided by the manufacturer himself.

Taken as a whole, cam grinding by any of the methods described is a compartively simple operation, once the operator thoroughly understands the fundamental principles of the work. This is necessary, as no workman should grope along in the dark as it were; he should understand thoroughly every step of his work and the reasons for the different operations. It is impossible to set any hard and fast rules, as every rule generally has an exception, and local conditions have to be given consideration in many cases.

# INDEX

# INDEX

## INDEX

Spotted something that isn't quite right?

Let us know and we'll fix it for future reprints. And as a thank you for your diligence, we'll send you a free Sothis book of your choice.

publisher@sothis.press

This page intentionally left blank

Made in United States
Orlando, FL
06 November 2026

90301613R00081